HIGH TECH
HARVEST

HIGH TECH
HARVEST

Understanding Genetically
Modified Food Plants

Paul F. Lurquin

Westview
PRESS

A Member of the Perseus Books Group

Copyright © 2002 by Westview Press, A Member of the Perseus Books Group

Westview Press books are available at special discounts for bulk purchases in the United States by corporations, institutions, and other organizations. For more information, please contact the Special Markets Department at the Perseus Books Group, 11 Cambridge Center, Cambridge MA 02142, or call (617) 252-5298.

Published in 2002 in the United States of America by Westview Press, 5500 Central Avenue, Boulder, Colorado 80301-2877, and in the United Kingdom by Westview Press, 12 Hid's Copse Road, Cumnor Hill, Oxford OX2 9JJ

Find us on the World Wide Web at www.westviewpress.com

Library of Congress Cataloging-in-Publication Data
Lurquin, Paul F.
  High tech harvest : understanding genetically modified food plants / Paul F. Lurquin.
    p. cm.
  Includes bibliographical references (p.).
  ISBN 0–8133–3946–4 (alk. paper)
  1. Transgenic plants. 2. Plant genetic engineering. 3. Plant biotechnology. I. Title.
SB123.57 .L88 2002
631.5'233—dc21

                                                          2002004938
The paper used in this publication meets the requirements of the American National Standard for Permanence of Paper for Printed Library Materials Z39.48-1984.

10   9   8   7   6   5   4   3   2   1

To Antigone, Jean-Paul,
Lev, and Louis-Ferdinand,
for general inspiration and attitudes in life.

Contents

List of Illustrations

Photos

Preface

GENETICALLY ENGINEERED PLANT PRODUCTS line the shelves of our grocery stores but we do not know which ones they are because no labels identify them. Should we be concerned? Should we—figuratively speaking—be up in arms against biotechnology as are the Europeans, the Japanese, and the Canadians? What are genetic engineering and biotechnology anyway? How does one genetically engineer plants? Is it true that some plants have been engineered with a gene extracted from a fish? Biotechnology companies are saying that engineered corn and canola are safe. Are they telling us the truth? I have written this book to answer all these questions and more.

I believe the public has the right to know and understand how its food is being manipulated at the most intimate level, that of the DNA itself. My goal is to inform, not to take a stand in favor or against genetically modified organisms (GMOs). I am, however, critical of the way biotech companies have introduced (or rather, failed to introduce) their plant products to the public. I am equally critical of those who show their disapproval of GMOs through acts of vandalism rather than with open discussion. This being said, I add that I have never received grant money from any biotech company, nor have I ever consulted for monetary gains for one. I do know plant genetic engineering quite well, however. I have been involved since 1973 in the basic research that led to its development, always in an academic environment. I have met most of the scientists who made plant genetic engineering possible, and as far as I know the vast majority of them are not only still alive but also still actively involved with their work. This good news does not attest to their longevity or mine, it simply shows that biotechnology is an extremely young science. We should keep this fact in mind when we think about its implications.

Biotechnology is an innovation that requires some explanation. It involves a type of genetic manipulation that is entirely new, but it relies squarely on fundamental scientific discoveries made in decades past. Many of these discoveries are complex even though their applications may seem deceptively simple. Therefore, one must understand genetic engineering before one can formulate an informed opinion about it. To become an informed person, one must do more than gloss over difficult concepts and then declare that one is for or against genetically modified foods. For this reason, parts of this book will require attentive reading, in particular Chapter 3, which explains the basics of gene cloning. The reader will then understand that biotechnology is an offshoot of the basic science of genetics, not a technology that was developed for the specific purpose of making genetically modified organisms. I hope also to demonstrate that in the end acceptance or rejection of genetically modified food plants must rely on science and science alone. Politics, economics, or other societal factors cannot replace the objective tools of the scientific method, whose validity has never been refuted successfully since its inception about 400 years ago.

This book originated with another book project I undertook with Columbia University Press in 1999. That book, *The Green Phoenix: A History of Genetically Modified Plants*, published in 2001, is a scholarly work intended mostly for academics and university students. Holly Hodder (former publisher for the sciences at Columbia University Press and now vice president and publisher at Westview Press) then suggested that a general audience trade book on the history and applications of plant genetic engineering would serve a purpose—that of informing the public of what is happening in this field. Hence, this book. Basic genetic principles and elements of gene cloning are presented before plant genetic engineering proper and its implications. My philosophy here is simply that the cart should not precede the horse. Too many reports have assumed that readers already know genetics, and this assumption has resulted in the hideously wrong—but fairly common—misconception that anything that contains genes is, by definition, bad. All living creatures have genes, and we have learned to manipulate these with ultimate precision. That is a great and perhaps frightening novelty. After reading this book, the reader will understand how this knowledge and power evolved.

Hundreds of scientific articles dealing with plant genetic engineering have been published. Some are listed in the references at the end of this book. This bibliography presents articles that do not require any particular scientific knowledge in order to understand them, such as *Scientific American* articles, along with some primary sources that do require deeper knowledge of biology. This book, however, does not require any advanced understanding of biological science. In addition to literature references, relevant web sites, both for and against plant biotechnology, are also provided.

I am grateful to Jerry Swensen, Lászlo Márton, Charlotte Omoto, and Diter von Wettstein for reading drafts of this work and pointing out places where clarifications were needed. I also thank them for expressing their own viewpoints regarding plant biotechnology in general. I am particularly indebted to my wife, Linda Stone, for her careful multiple readings of the manuscript. Last but not least, my deepest gratitude goes to Holly Hodder, my editor, and Catherine Hope, my copy editor, for their meticulous editing of the manuscript and excellent stylistic suggestions. As usual, all errors, interpretations, and omissions are mine. Finally, I hope readers find this book a useful tool. Public opinion of plant biotechnology has become an emotional morass. I offer

this book as a way for people to inform themselves and make up
their minds in an objective way. This end is, perhaps, the best a
scientist can hope to achieve.

Paul Lurquin
Pullman, Washington and
Cannon Beach, Oregon

Chronology of Events Described in This Book

Technical terms are defined in the glossary and in the main text.

1865 Gregor Mendel discovers genes and the laws of heredity.

1907 E. F. Smith and C. O. Townsend discover that crown gall tumors in plants are induced by the bacterium *Agrobacterium tumefaciens*.

1910 Thomas Morgan demonstrates that genes are on chromosomes.

1944 Oswald Avery and collaborators demonstrate that DNA is the material of which genes are made and discover transformation through DNA uptake.

1953 James Watson, Francis Crick, Rosalind Franklin, and Maurice Wilkins determine the double helical structure of DNA.

1953 William Hayes establishes the concept of plasmid DNA.

1962 Werner Arber and his group discover bacterial restriction endonucleases.

1966 Marshall Nierenberg and Gobind Khorana finish deciphering the genetic code.

1967 Jerome Vinograd and collaborators invent a technique to isolate and purify plasmid DNA.

1968 First experiments aimed at investigating DNA uptake in plants are conducted.

1970 M. Mandel and A. Higa develop transformation of the bacterium *Escherichia coli*.

1970 Georges Morel proposes that crown gall tumors appear on plants as the result of genetic information transfer from *Agrobacterium tumefaciens* to plant cells.

1972 Herbert Boyer and Stanley Cohen perform the first cloning experiment with plasmids.

1972 First attempts to produce genetic effects in plants with externally supplied foreign DNA are undertaken.

1974 Jef Schell, Marc Van Montagu, and others discover large plasmids in virulent *Agrobacterium tumefaciens*.

1976 Mary-Dell Chilton, Eugene Nester, Milton Gordon, and others discover gene transfer from *Agrobacterium tumefaciens* to plants.

1977 Walter Gilbert and Frederick Sanger develop techniques to determine the base sequence of DNA.

1982 Cell electroporation in the presence of DNA is invented.

1983 Jef Schell, Mary-Dell Chilton, Marc Van Montagu, Robert Fraley, Robert Horsch, and others transform plants with foreign genes via *Agrobacterium*-mediated gene transfer.

1984 Ingo Potrykus's group demonstrates plant transformation with naked recombinant DNA.

1987 The "gene gun" is invented.

1987 First demonstration that transgenic plants containing the *Bacillus thuringiensis* (Bt) toxin gene are resistant to certain insects takes place.

1987 Plant Genetic Systems generates plants resistant to the herbicide Liberty®.

1988 The first commercially available genetically engineered fruit, the FlavrSavr® tomato, is produced by Calgene.

1988 Monsanto generates soybean plants resistant to the herbicide Roundup®.

1995 First laboratory production of "plantigens" is under way.

1996 Massive transgenic crop sales begin.

1998 Prince Charles of Wales publicly declares his opposition to biotechnology.

1999 Of the total 72 million acres planted with soybeans in the United States, half were planted with Roundup®-resistant seeds.
1999 Protest against the use of genetically modified plants in foods is in full swing in the United States and Europe and leads to street demonstrations.
2000 Provitamin A-producing "golden rice" variety is created. Genetic engineering techniques now exist for just about every conceivable cultivated plant species, from apple trees to coffee, from bananas to asparagus, to eggplant, to lettuce, to wheat.

Introduction:
The Old and the New

I N A SENSE, HUMANS *DO* LIVE by bread alone. Practically all life on Earth, animal and human, ultimately depends on the ability of plants to capture the photons of light released by our star, the Sun. The only well-documented exceptions are some microbial and worm communities that dwell in complete darkness near hydrothermal vents located deep under water on the ocean floor. These communities are sustained by chemical reactions taking place in the superheated water spewing out of these vents. Yet, even some of these creatures depend on oxygen dissolved in sea water; and their oxygen, like ours, is produced by plant life.

Plants, from microscopic marine phytoplankton to majestic sequoias as well as humble domesticated species, use sunlight to split water molecules into breathable oxygen and hydrogen ions (protons) and electrons. Oxygen is released into the atmosphere, and protons and electrons are used to power the reactions that reduce atmospheric carbon dioxide into sugars. These sugars are in turn metabolized through mechanisms that result in plant cell growth and development. Freed oxygen is used further by plants themselves and all animal species for crucial metabolic processes, survival, and proliferation.

Photosynthesis, as this light-harvesting mechanism is known, appeared approximately 3.5 billion years ago, roughly 1 billion

1

years after the Earth formed. Microscopic bacterial cells (cyanobacteria), not land plants, first developed photosynthesis based on chlorophyll. Cyanobacteria still exist today and are everywhere. The first land plants appeared a little over 400 million years ago during the Silurian period of the Paleozoic era and were accompanied or closely followed by the first land animals. Much, much later, about 5 million years ago, our earliest-known bipedal ancestor, *Australopithecus,* roamed the African savanna in search of edible plants. Thus, for about 400 million years, plant life was left undisturbed except by natural events such as mutations, fires, and changing climate patterns that drove the slow process of evolution and led to great diversity of form. Evolution is driven by natural selection: Naturally occurring mutations can be favored or not by certain ecological conditions. This process can lead to the proliferation or extinction of species, both perfectly natural occurrences.

Then, about 10,000 years ago, *Homo sapiens,* ourselves, completely changed plant evolution by replacing natural selection with artificial, directed selection in some plant species. Today we call this agriculture. Therefore, genetic manipulation of plants is not new. The beginnings were modest; progenitors of modern wheat were first domesticated from wild relatives in the Middle East, whereas ancestors of modern corn appeared later in Mesoamerica. In this way, humans created the first engineered plants, possibly to free themselves from the vagaries of hunting and gathering in some ecological settings. (Another theory states that plant domestication was innovative but not caused by a need for more food. For the purpose of this book, regardless of which theory is more accurate, the result is the same.) Agriculture was soon followed by the establishment of cities, social classes, taxation, and writing, and it remains vital today.

The first experiments in plant breeding (along with animal husbandry) must have been performed purely by trial and error until a desirable result was obtained. It was not until the second half of the nineteenth century that plant breeding was put on a scientific basis by Gregor Mendel, the father of genetics. In the decades since that time, humans have learned how to alter the hereditary fabric of plants at the deepest level, that of the DNA molecule. We call this *plant biotechnology*, and the results of this technology are genetically modified (also called transgenic) plants.

It took a mere 10,000 years or so for humans to take control over the most intimate mechanisms of plant life and reproduction. What took evolution by natural selection millions of years to achieve, we can now alter in just a few weeks. Moreover, biotechnology can now do what nature could not; it can blend genes from totally different species, genera, and kingdoms of life. Biotechnology can defeat sexual barriers.

As just noted, humans had learned to manipulate plant genes in a crude manner well before the invention of biotechnology. How did this come about? Surely, cultivated crop plants look very different from their wild relatives. A classic example is that of teosinte, the ancestor of corn, that has very small, spiky ears and tiny seeds. Mesoamerican people, perhaps as early as 7,000 years ago, domesticated teosinte by finding spontaneous variants, or mutants, devoid of these undesirable characteristics and then propagated them. Wheat first appeared in the Fertile Crescent of the Middle East and resulted from accidental or deliberate hybridization between two related grass species that by themselves were far less convenient to harvest and process.

Ancient and not-so-ancient examples of stable plant hybrids and mutants created or propagated by humans are numerous and include cotton, chrysanthemum, potatoes, bananas, seedless grapes and watermelons, tiger lilies, some apple varieties (Winesap and McIntosh, for example), coffee, alfalfa, peanuts, triticale, strawberries, and some petunia varieties. In all these cases, either the progenitors of these new varieties are sexually compatible (they can fertilize one another) or their chromosome number has changed, which often happens accidentally and naturally. Nevertheless, without human intervention in the form of selection for useful traits and massive propagation, these plant varieties would be rarities. In that sense, humans have manipulated plant genomes (the genome is the suite of all genes contained in an organism) for thousands of years without any knowledge of genes or DNA. Some have called this *primitive biotechnology*.

Yet modern biotechnology is different. Here, theoretically, any plant species can be modified with genes from any source: bacterial, animal, fungal, or another plant. This technology no longer relies on natural hybrid formation or accidental changes in chromosome numbers. Rather, it hinges on our ability to isolate and clone genes from any species and introduce these cloned genes into plant cells through a variety of techniques. In other words, plants genetically modified by cloned foreign genes cannot be produced by nature. Scientists have now created plants that produce human proteins, express bacterial or fish genes, make plastics, and are able to detoxify toxic wastes. Genetically modified plants can do essentially whatever scientists force them to do. To use an old phrase, we can indeed play God.

This book tells the story of plant genetic manipulation by humans. It starts with the discoveries of Mendel and other early pioneers, which were followed many years later by the recombinant DNA revolution and demonstration that the genetic engineering of plants is possible. This story is not simple, and, in my opinion, it is impossible to have a good grasp of biotechnology without understanding genetics. In turn, it is impossible to understand genetics without comprehension of the gene as Mendel defined it well over 100 years ago. His discoveries are not obsolete, and it would be a mistake to believe that the modern DNA gene is somehow more real than the old Mendelian gene. Both are different facets of the same reality. Just as rocket scientists use Newton's laws of gravitation, published in 1687, to launch spacecraft that can reach the outer planets and beyond, so do biotechnologists use Mendel's laws of inheritance, published in 1865, to verify that their genetically modified plants do what they are supposed to do. After discussing Mendelian genetics, I will explain what genes are made of, how they work, and how this knowledge was acquired. This material constitutes Chapters 1 and 2. Current DNA cloning techniques will then be covered in Chapter 3. This overview of classical (Mendelian) and molecular (DNA-based) genetics will thus set the stage for the study of plant biotechnology proper, beginning with Chapter 4.

We will then see what commercial and basic applications were derived from the new technology and discuss the societal and economic consequences of these applications (Chapters 5 through 7). These are emotion-laden issues that have triggered street demonstrations, inflammatory declarations by celebrities, and commercial negotiations at the international level. It goes without saying that this clamor is not just political hype; some of the concerns about plant biotechnology are definitely valid. Unfortunately, the science behind this technology as well as its positive applications have not been cogently communicated to the public. A main goal of this book is to redress this situation and help the reader distinguish fact from fiction.

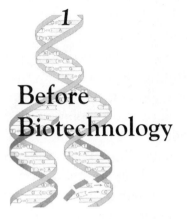

Before Biotechnology

O UR PRESENT-DAY ABILITY TO MANIPULATE genes did not come overnight, as a great flash of creativity striking a single savant. This capability is the result of work done by thousands of scientists who have assembled knowledge on a foundation of a few landmark discoveries. These discoveries will be explained in this chapter and in the next two chapters. It should be realized first that biotechnology is a very young science, itself a part of the young science of genetics. In fact, my own grandparents were little children when Gregor Mendel, the inventor of genetics as we know it today, was still alive. The nature of the genetic material DNA was discovered when I was two years old, and I was in elementary school when the DNA double helix was revealed by James Watson and Francis Crick. The first cloned DNA molecules were produced in the early 1970s, shortly after I received my Ph.D., and the first genetically engineered plants were created in 1983, in other words, yesterday. Commercial exploitation of genetically modified plants started only a few years before the end of the last millennium. Biotechnology is baby-boomers' science: Its story is ours.

Before Genetics

Our ancestors must have realized a very long time ago that like begets like. They certainly noticed that corn seeds always germinated to give corn plants, which in turn gave more corn seeds, and so forth. Similarly, when their dogs mated, their offspring were more dogs. They must also have noticed that sometimes, however, the progeny was not quite exactly identical to the parents. Once in a while, a corn ear would bear some pale yellow kernels in the midst of dark blue ones, the original color of corn kernels. These were mutants. Similarly, coat color in dogs undoubtedly varied among the individuals of the same litter. This departure was due to gene reassortment. There is no evidence that our forebears understood this underlying reason, however. Rare mutation events must have seemed ominous: For example, occasional human albino mutants in ancient societies were the victims of sacrifice or, alternatively, deification.

Humans must have found it very strange indeed that offspring could sometimes look sharply different in some ways from the parents—plant, animal, or human. The idea of the blending of parental traits in their offspring, as was commonly accepted for eons, contradicted this result. After all, in humans the progeny of a black parent and a white parent has intermediate skin color, showing no drastically sharp difference from the mother and father. This characteristic makes perfect sense if one believes that inheritance results from the blending of the parents' visible characteristics (called the *phenotype*) and produces physical traits that are "in between." Yet, why is it that two perfectly normally pigmented parents can give birth to an albino child? There is no blending of traits here. Further, why is it that the children of an albino and a normal parent almost invariably show regular pigmentation? Again, no blending. Clearly, blending inheritance does not always apply. Similarly, ornamental plant breeders must have noticed at one point or another that a hybrid between a red flowering plant and a white flowering one gave offspring with red flowers, not pink ones. Yet, well into the nineteenth century, blending inheritance was not questioned. It was eventually the work of the great hermit scientist Mendel that explained all these conflicting observations, made sense of heredity, and created the science of genetics.

The Discovery of Genes

Gregor Mendel (1822–1884) was the first person to understand that the theory of blending inheritance was wrong and that genes behave as discrete units, passing from parents to offspring in completely predictable ways. Most people know that he was a monk, born in Moravia, then part of the Austro-Hungarian empire and now a province of the Czech Republic, and that he hybridized pea plants. What most people do not know is that he had been trained in physics and was the first scientist to apply quantitative thinking to a biological problem. By *quantitative*, I mean that Mendel actually tallied up the plants he used and produced in his experiments, rather than simply looking at them, which was the norm at the time in the life sciences. Mendel did not use any scientific instruments other than a simple microscope for his experiments, and he never ground up his plants to see what they were made of. His most important tool was his brain. Also, he did not know about DNA, which had been discovered during his lifetime but whose function remained mysterious until 1944 and whose structure was unknown until 1953. Moreover, he did not even know about meiosis, the special cell division that generates germ cells—egg and sperm—and which was discovered the year before his death. Yet, he invented genetics.

How did he do it? First, he was a genius, and second, because he was a monk (although not a celibate one; apparently he fathered a son), he must have had plenty of time available to do his seemingly useless research on pea plants. He must also have been driven by pure scientific curiosity, since there is no evidence that the monks of Mendel's monastery sold vegetables for a living and so needed better-quality peas. It is noteworthy, however, that Mendel produced a pea variety so tasty that it was used for years in his monastery's kitchen. Further, Mendel was a high school teacher and a well-known breeder of ornamental fuchsia plants. But first and foremost, he was a true, critically minded scientist who combined in a positive way outstanding basic and applied research.

It is perhaps best to digress a little and take a look at the scientific process itself. In a perfect world, scientists first become fully acquainted with their field of research by reading the scientific literature. Next, loaded with this baggage and aware of all the natural phenomena in need of an explanation, they formulate hypotheses (basically ideas, sometimes even considered "off the wall") to rationalize the unknowns. These hypotheses must then be tested by controlled experiments, after which they will be accepted or rejected, depending on the results of the experiments. A verified hypothesis usually leads to more hypotheses, which must in turn be tested, and so forth. A set of verified hypotheses can then become a theory. A theory is a group of verified hypotheses with general applicability to a group of natural phenomena. Finally, a scientific law is a theory that seems to be entirely universal, with no known exceptions. One well-known example is the law of gravitation, which explains how planets orbit around the Sun and how galaxies orbit around one another. Other more obscure examples are the laws of thermodynamics that regulate energy changes in physical systems.

It so happens that the whole of biology contains only a single set of laws: Mendel's. No other theory merits the designation of law in the life sciences.[1] This attests to the importance of Mendel's discoveries. Did Mendel actually formulate one or several hypotheses regarding the nature of heredity before starting his experimental work? Or did he first experiment with his pea plants and only afterward try to make sense of what he observed? In his published work, Mendel did state that he was aware of experiments done by other breeders, but he did not divulge his own thought process. We can only speculate on the reasons that drove him to conduct the experiments that I will describe. This description of his work may seem overly detailed, but one must realize that what is being outlined here is no less than the birth of genetics.

Mendel had at his disposal fourteen lines of peas (*Pisum sativum*) that differed by sharply distinguishable characters. These fourteen lines corresponded to the following seven pairs of contrasting traits:

flower color (with one variant bearing purple flowers and the other white flowers)
flower position (lateral or terminal)
seed color (yellow or green)
seed shape (round or wrinkled)
pod shape (full or constricted)
pod color (green or yellow)
stem length (tall or dwarf)

He decided to do hybridization experiments with pea plants each of which *differed by a single character* such as flower color, seed shape, and so on. But first, let me describe what a hybridization experiment is.

Peas are self-fertilizing, meaning that the pollen grains (male germ cells) carried by a flower will fertilize the ovules (female germ cells) carried by the same flower. This is because the male and female organs of a pea flower are in close contact during flower development. However, the large size of the flowers makes cross-fertilization by human intervention possible. In this case, the experimenter snips away with a pair of fine scissors the pollen-carrying anthers before self-fertilization has occurred. Then, using a fine brush, the experimenter can rub the pistil (the female sex organ) of the castrated flower with pollen harvested from another plant. For example, one could fertilize the ovules (in plants, the equivalent of eggs) of a white flower with pollen from a purple flower and vice versa. After self-fertilization or cross-fertilization, the plants are allowed to set seed, and these seeds are then planted, allowed to grow, and examined for expected or unexpected traits (phenotypes). Plants from which the pollen grains and ovules originally derive are called the parental generation, while the plants resulting from self- and cross-fertilization events are called the first filial, or F1, generation. Self-fertilized F1 plants will give rise to the second filial, or F2, generation.

Mendel's first procedure was to ensure that his fourteen pea lines bred true. That is, he verified that his purple flower-bearing line, for example, always produced purple flowers over several generations and not occasionally white or other colored flowers. This meant that the traits he was observing were stable and that his seed stocks were uncontaminated. Next, he cross-fertilized plants that differed by a single character. For example, he took pollen from a purple flower and fertilized with it the ovules of a white flower. He did the same experiment with plants carrying round seeds versus wrinkled seeds and continued with the remaining five pairs of contrasting characters. For brevity, we will only look at the results he obtained with the purple/white flowering pair as the parental generation. Mendel realized that the F1 hybrid generation originating from a cross-fertilization event of these parents carried 100 percent purple flowers. Everything looked as if the white trait had disappeared, and there was no blending of traits either! But was it really true that the white trait had disappeared? To determine this, he let these F1 plants self-fertilize and looked at flower color in the second filial generation, the F2. And there, to his surprise, white-flowering plants reappeared! The white trait had not disappeared in the F1, but was simply masked by the purple trait. We now say that purple is *dominant* and that white is *recessive*.

However, Mendel did not stop after naming those traits. He actually counted the number of F2 purple-flowering plants and compared that number to the number of F2 white-flowering plants. He had examined exactly 929 F2 plants, of which 705 turned out purple and 224 were white. Thus, 75.89 percent of the

F2 generation were purple and 24.11 percent were white. (I am intentionally giving these numbers with two decimals rather than rounding them; see the explanation that follows.) Practically identical results were obtained in the mating crosses involving the other six pairs of characters. In other words, in F2, the dominant trait was roughly three times more numerous than the recessive trait. It is at this point that Mendel must have realized that he had discovered an important property of heredity: In the F2 generation, the ratios between dominant and recessive individuals are always very close to 3:1. It may have been at this point that Mendel's background in physics helped him conceptualize his results and put them into a theoretical framework. Mendel posited that his observed ratios were *not just very close* to 3:1, they *represented exactly* 3:1 ratios or, in the example above, *exactly* 75 percent and 25 percent. He attributed the fluctuation around his 3:1 ratios to statistical imprecision due to limited sample size. We know today that he was right; as the numbers of examined F2 plants increase, the actual ratios get closer and closer to an ideal 3:1, or an ideal 75 percent and 25 percent.

Mendel needed to explain the reasons why this 3:1 ratio always cropped up in his crosses involving a pair of contrasting characters; in this step is where his genius truly showed. He realized that his results could be explained only if the following were true:

1. Physical traits such as flower color, seed shape, and so on are determined by discrete, indivisible units of heredity. (These units are now called *genes*.)
2. A single gene, such as the one determining flower color in pea plants, can exist in the form of several variants, one that directs white color and another that directs purple color. (The variants of the same gene are now called *alleles*.) As he found out, some alleles are dominant over others, purple being dominant over recessive white in peas.
3. Plants contain two alleles for each gene. If the two alleles are identical, two copies of the purple allele or two copies of the white allele, the plant is said to be *homozygous*. If the two alleles are different, one copy of purple and one copy of white, the

plant is said to be *heterozygous*. Since purple is dominant over white, the heterozygote will be purple, as in the F1 generation of the previously described cross.

4. When pea plants reproduce through the formation of male and female germ cells (called *gametes*), the two alleles must separate, and only a single copy of an allele finds itself in each gamete. When the male gamete, carrying one allele, fertilizes an ovule (an egg), also carrying a single allele, the resulting offspring carries again two alleles. If the alleles carried by the pollen grain and the ovule are the same, the resulting offspring is homozygous. If the alleles carried by the pollen grain and the ovule are different (for example, a purple allele in the pollen grain and a white allele in the ovule), the resulting plant is heterozygous. Thus, traits, technically called phenotypes, are under the control of *pairs* of alleles. The alleles themselves constitute what is called the genotype. The separation of alleles during gamete formation is now called the *law of segregation,* or *Mendel's first law.* We know today that physical separation of alleles during gamete formation is due to the fact that each allele is carried by a different chromosome and that these chromosomes physically separate during meiosis, the special type of cell division responsible for gamete formation. To fully appreciate Mendel's conclusions, we must remember that meiosis was unknown in his time.

We now know that Mendel's inferences were right. The *only* way to explain the results of his hybridization experiments was to make the assumptions described above. Absolutely no other explanation can account for his observations and all other observations made by countless experimenters after him.

Many people find Mendel's laws abstruse, dry, and difficult to digest. This is the case in part because the sophistication of the experiments he did was not up to par with the intellectual quantum leap of their interpretation. After all, anybody can hybridize peas, but few people would be able to draw Mendel's revolutionary conclusions from such seemingly mundane gardening practices. And let's face it, these concepts are difficult because they are abstract. Yet, Mendel's discovery impacts our daily lives, from the consumption of genetically modified corn flakes to the calculation of the probability of giving birth to a genetically challenged child.

A simple diagram helps clarify Mendel's ideas. For this, we will use a Punnett square, named after an early-twentieth-century geneticist who designed this simplifying aid. In our example, using purple and white flowers, we will use the symbol P for dominant purple and p for recessive white (recessive alleles are often the lower case version of the dominant allele). This can be confusing. Why not call purple "P" and white "w"? Well, this would result in even more confusion. We can thus write the crosses like this:

The parental generation (original homozygous, true-breeding parents) is purple for one parent and white for the other. Thus, we have

	Purple	x	White	(cross fertilization event)
That is	PP	x	pp	(using Mendel's idea that alleles come in pairs)

This cross gives in F1 100 percent purple individuals of heterozygous genotype Pp, since a single P allele was inherited from the PP parent and a single p allele was inherited from the pp parent. F1 hybrid individuals are then allowed to self-fertilize, which can be written

	Purple	x	Purple	(self-fertilization event)
That is	Pp	x	Pp	

To determine the outcome of this cross and obtain the genotypes of the F2 generation, one must first determine what gametes (carrying a single allele) these hybrids are going to produce. According to Mendel's first law, the P and p alleles will segregate (separate) during formation of the germ cells and each F1 individual will thus produce 50 percent P gametes and 50 percent p gametes. We can insert these symbols into a Punnett square (below) and simply combine rows and columns to recreate the genotypes of the F2 individuals.

Male Gametes

		P	p
	P	PP	Pp
Female Gametes			
	p	Pp	pp

Note that the P and p alleles are represented equally (50 percent each) in both the male and female parental gametes. Each of the four boxes of the Punnett square contains one possible genotype, which, taken all together, represent the whole F2 population, the offspring of the parents. These genotypes are: PP, Pp (twice), and pp. But what are the phenotypes, given that P is dominant over p? The PP individuals will of course be purple, and so will be the Pp heterozygotes. The pp individuals, being homozygous recessive, will of course be white. Therefore, the ratios are 3 purple for 1 white, that is 3:1, exactly what was observed by Mendel and explained by his law of segregation.

The validity of Mendel's interpretation of heredity has been verified countless times and applies to myriad life forms, including humans. We will not concern ourselves with Mendel's second law, the law of independent assortment,[2] as it is not important for our ultimate purpose, the explanation of the making of genetically modified plants. These laws not only set the foundation of the science of genetics, they are still used universally by breeders, genetic engineers, and genetic counselors alike. In fact, Mendel's laws are so generally applicable that plant genetic engineers use them today to determine whether a newly inserted foreign gene is present in a homozygous or heterozygous state and how many copies of this foreign gene are present in the engineered plants. Likewise, genetic counselors can tell concerned parents what the probabilities are that their unborn child will suffer from a hereditary disease. Mendel's laws have absolute predictive value and can be applied as surely as the laws of physics and chemistry. We will see in Chapter 4 that several investigators claimed to have genetically modified plants with externally supplied DNA as early as the mid-1970s. These claims were in error, and this mishap was due in large part to the refusal of these authors to realize that their results did not conform to Mendel's first law.

Going back to the example of skin color in humans, we now know that skin color is a polygenic trait, that is, a phenotype under the control of many genes. Blending is understood in terms of these many genes, each contributing a fraction to pigment formation. These genes, however, still obey Mendel's laws. In contrast, albino individuals make no pigments at all because the mutation of a "master" gene (I simplify here) has rendered completely inoperative their pigment synthetic pathway. Their offspring are normal because the gene of the other parent is normal, compensates for the effect (absence of pigmentation) of the defective gene, and is dominant. The albino gene (or allele, which is more correct) is recessive and has no influence on the phenotype of the offspring, provided that a functional copy of that gene, the dominant allele, is contributed by the other parent. Likewise, these are Mendelian genes.

Although Mendel discovered the concept of the gene and the laws that dictate their transmission from parents to progeny, that is all it was then—a concept without physical basis. It would take another eighty years or so to discover from what genes are made. Meanwhile, Mendel's discovery was mostly ignored by his contemporaries and ultimately forgotten for more than three decades; he had been preaching in the desert. Historians blame this loss on a world that was not ready to understand Mendelian laws of heredity. As we have seen, Mendel provided very abstract mathematical explanations for the behavior of traits in genetic crosses. Biologists of his day were completely descriptive and far from working with unifying theories regarding the complicated science of life. There was one great exception, however: Charles Darwin (1809–1882). He was a contemporary of Mendel and (with his competitor Alfred Wallace) had by then developed the theory of evolution by natural selection. Darwin, who was in need of an explanation for the transmission of new traits brought about by evolutionary processes, never resorted to Mendel's laws as a framework for his own theory. This omission was perhaps caused by Mendel himself, who had stated that it was necessary to understand the formation of germ cells and the process of fertilization before his laws of inheritance could be generally accepted. The synthesis of the two great principles, the gene and evolution by natural selection, was finally achieved several decades after both authors' deaths.

Mendel's laws were rediscovered at the turn of the twentieth century by three botanists, Carl Correns of Germany, Hugo de Vries of the Netherlands, and Erich von Tschermak of Austria. From then on, classical genetics, that is, the study of genes as entities that determine phenotypes, without consideration of their molecular nature, was on its way. However, one obvious question was, Where in a living cell are these genes located? It was soon proposed that chromosomes were the physical supports of Mendel's genes. The chromosomal hypothesis was formulated in 1903, almost simultaneously, by two geneticists, Walter Sutton of the United States and Theodor Boveri of Germany.

What are chromosomes? They are elongated strandlike bodies, easily distinguished under a simple microscope, that are visible only in cells undergoing division. In nondividing cells, chromosomes are bunched up in the cell nucleus and cannot be distinguished from one another. Chromosomes, so named because they are stainable by a variety of dyes, were seen first to double in number and then to divide equally between daughter cells during division. Since all descendants of a cell have the same genetic properties as the parent cell, it made sense to attribute the location of genes to chromosomes. These chromosomal bodies first duplicated (the number of genes in that cell was doubled) and then separated equally during division (the doubled number of genes was halved), thus restoring the original number of genes to both daughter cells, making them identical to the parental cell. What is more, meiosis, the cell division process that produces gametes, was found to produce cells that contained half the number of chromosomes of regular cells. This result was exactly what Mendel had predicted. And again, it made sense. Let us look at an organism such as a human, whose cells contain two copies of every gene and chromosome. During meiosis, sperm cells and eggs are produced that contain only one copy of each gene and chromosome. When a sperm and an egg combine to produce a human zygote (a single-celled embryo), the number of chromosomes and gene copies will be restored to their original value, two of each, one from the mother and one from the father. Organisms that contain two copies of each gene and chromosome, such as humans, barley, alligators, and lions, are called *diploid*. Their gametes are said to be *haploid*, meaning that they contain only one copy of each gene and chromosome. And indeed, what would happen if gene and chromosome numbers were not halved during meiosis? The father would contribute two copies of each gene and chromosome and so would the mother. The unfortunate offspring would then hold four copies of each gene and chromosome. This number would of course increase to eight in the next generation, and so on. One sees the absurdity of this situation because at a certain point, cells would not be big enough to host all these chromosomes.

Genes Are Located on Chromosomes

The chromosome theory (or hypothesis, at that point) of genes was put on solid and definitive ground by Thomas Hunt Morgan and his students, most notably Calvin Bridges and Alfred Sturtevant, in their laboratory at Columbia University in the second decade of the twentieth century. Morgan's laboratory veered away from plants as objects of study and inaugurated the fruit fly *Drosophila melanogaster* as a model for genetic studies. *Drosophila* is still much in use today, and the DNA sequence of this organism was published in March 2000 in an article coauthored by no fewer than 205 people. "Big science" is no longer restricted to atom-smashing physicists. This DNA sequence has revealed that *Drosophila* and humans share many genes in common, making this fly an interesting model for the study of disease genes in people.

Morgan and his collaborators were able to prove that genes are indeed associated with chromosomes upon which they are arranged in a linear order. To accomplish this task, they took advantage of the fact that like humans, female *Drosophila* has two copies of the sex-determining X chromosome and is genotypically XX (double X), yet male *Drosophila* has only one X chromosome and one Y chromosome, which in this organism determines male fertility. Males are thus genotypically XY. What is more, the Y chromosome does not carry alleles of genes present on the X chromosome. This means that in males, a recessive trait associated with the X chromosome will always be expressed, since the Y chromosome possesses neither recessive nor dominant alleles of the genes carried by X. In addition to understanding this, Morgan had isolated a white-eyed *Drosophila* mutant (normal eye color is red), and this allele proved in crosses (matings) to be recessive to red.

A series of crosses between the mutant individuals and normal flies demonstrated that the pattern of inheritance of the white-eye allele could only be explained if this allele was located on the X chromosome. This conclusion was made possible because of the different chromosome combinations carried by the two sexes, XX and XY. By definition, sons receive their Y chromosome from the father and their X chromosome from the mother. Morgan observed that a white-eyed female (thus homozygous for the mutant allele) crossed with a normal male always produced white-eyed sons. Conversely, homozygous red-eyed females crossed with a mutant white-eyed male always produced red-eyed sons. In other words, the gene for eye color followed the inheritance pattern of the X chromosome. A cross between a white-eyed female and a red-eyed male is represented in the Punnett square below and illustrates one of Morgan's experiments. Using again the counterintuitive genetic nomenclature, we will call the allele determining red eye color W and the allele determining white eye color w. W is dominant over w. An X chromosome carrying a red allele will thus be designated X^W while an X chromosome carrying a white allele will be designated X^w. The Y chromosome does not carry alleles for eye color.

Thus, we do the following cross,

X^wX^w (white-eyed female) x X^WY(red-eyed male)

and we build a Punnett square with the male and female parental gametes:

	Female	
	X^w	X^w
X^W	X^WX^w	X^WX^w
Y	X^wY	X^wY

Male is the label for the left (row) gametes (X^W and Y).

It is clear that all the daughters (XX) are heterozygous for the W/w pair and are thus red-eyed, while the sons (XY), having inherited the w allele from the mother and a Y chromosome from the father, are white-eyed. You should feel free to verify this inheritance pattern by solving a cross involving a *homozygous* red-eyed female and a white-eyed male. Remember that the Y chromosome is inert in this cross too, as it does not carry any gene for eye color. The conclusion of these experiments clearly showed that the gene determining eye color in *Drosophila* is located on a chromosome, the X chromosome.

More X-associated genes were later discovered in Morgan's laboratory and used to determine their spacing on this chromosome. This could be done because chromosomes pair up during meiosis and, during this event, can break and become reunited with the other chromosome belonging to the same pair. The frequency of breakage and reunion depends on the distance between two locations on the chromosome. If the distance between two points is large, the number of breakage and reunion events (called crossover events) is large. The number of crossover events between two closely positioned points will of course be smaller. Ingenious crosses allowed the researchers to measure the distance between several X-linked genes, based on the frequency with which these genes crossed over, or, to use more modern phraseology, recombined. Morgan and his collaborators had thus demonstrated that genes are located on chromosomes and that the distances between genes could be measured. This methodology was then applied to genes located on all *Drosophila* chromosomes, not just the sex chromosomes. The same technique is in fact used today, without modification, by breeders and by genetic engineers who want to determine where the foreign gene they have inserted into their plants is located. Here again, technology more than eighty years old is fully applicable to modern problems. Morgan received the Nobel Prize in 1933, the first Nobel to go to a scientist studying heredity. There would be many more.

As just mentioned, chromosomes can be conveniently visualized under the microscope. It was soon realized that in many organisms, different chromosome pairs could be distinguished from their neighboring pairs by size and staining patterns. This branch of genetics that studies chromosomes microscopically is called *cytogenetics* and is used, among other ends, to detect chromosomal abnormalities in humans. These studies soon showed that meiosis does not always result in the perfect separation of chromosomes, and that sometimes, diploid gametes, instead of haploid gametes, can be formed by mistake. These gametes then contain a full chromosome complement (two copies of each). The union of a diploid gamete with a normal haploid gamete of the opposite sex will result in the production of a triploid individual, one that contains three sets of chromosomes instead of the regular two. This situation is not tolerated by mammalian zygotes, which soon abort. However, plants are much more flexible in their acceptance of supernumerary chromosomes and triploid individuals can survive.

You probably did not know that the seedless bananas you eat for breakfast are in fact triploid individuals. So are seedless watermelons. Wild bananas are diploid, full of seeds, and do not correspond to our idea of edible bananas. Yet the bananas we eat contain no seeds. Why are these fruits seedless, given that seeds are the equivalent of embryos in plants? The answer is that these triploid plants are sterile. Simply put, meiosis evolved to deal with even numbers of chromosomes, not odd numbers. Diploid organisms will produce haploid gametes (as is true in the enormous majority of plants and animals), tetraploid organisms (carrying four copies of the chromosome complement) will yield diploid gametes, and octaploid organisms will generate tetraploid gametes (containing four copies of each chromosome). A triploid organism, however, cannot subdivide its chromosome complement evenly, because there is no such thing as one and one half chromosomes. The same holds true for a pentaploid organism containing five sets of chromosomes; its gametes cannot harbor two and one half pairs of chromosomes. Thus, the mechanisms of meiosis get confused with organisms containing odd numbers of chromosome pairs and as a result, do not make viable gametes. In other words, triploid and pentaploid plants are sterile and form no seeds. These plants are thus not propagated through seed; they are multiplied through the cultivation of cuttings. The occasional seed found in a seedless watermelon or grape is due to

another infrequent mistake at the level of meiosis and the hap-
hazard production of rare haploid or diploid gametes.
Alternatively, commercial strawberries are octaploid (this condi-
tion makes the fruit bigger and more appealing) and make
gametes containing four complements of chromosomes, an even
number. These plants are fertile. So is modern wheat, a hexaploid
plant (six sets of chromosomes, an even number) that originated
from three different diploid progenitors. In the same vein, the
much favored Russett Burbank potato is a tetraploid tuber.

Humans have knowingly, and at times unknowingly, taken
advantage of nature's errors in meiosis to propagate food plants
that otherwise would have been evolutionary dead ends. Seedless
bananas, watermelons, oranges, and grapes cannot survive without human intervention. Most of this was made possible due to
the development of the science of classical genetics. The propagation and commercialization of these sterile and therefore evo-
lutionarily unfit species is, of course, an example of the early
application of genetic technology. The same holds true for fertile
plants such as corn and wheat. However, simple genetic technology always relied on the hybridization of sexually compatible
species (we now call this technique *classical breeding*) and, as we
have seen, chance events such as mutations that needed to be sin-
gled out. Modern biotechnology relies on very precise gene
manipulation and leaves very little room for chance. As we will
see in the next chapter, it is the discovery of the chemical nature
of the genetic material that made these refined genetic manipulations possible.

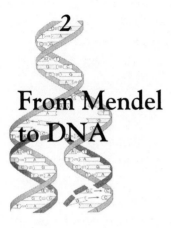

2

From Mendel to DNA

IMAGINE YOURSELF IN ENGLAND in 1953. Imagine further that you work in one of the laboratories of King's College in London. There, you often see a young scientist named Rosalind Franklin busy crystallizing DNA, the genetic material, and taking X-ray pictures of the crystals. Franklin is not very communicative, perhaps because hers is a male-dominated world of scientific research. She does not interact well with her supervisor and works essentially alone. Yet, she knows she has produced the best ever X-ray pictures of DNA, pictures that contain the secret of the structure of genes. Franklin is a stickler for quality and completeness. She does not want to publish her results in a rush and feels she needs a better understanding of their meaning before going public with her work.

Unfortunately, within a matter of weeks after her discovery, others will become acquainted with her precious, unique data, give them the correct interpretation, and eventually win the Nobel Prize. Franklin's efforts were barely acknowledged at the time by her competitors; she remains a largely unsung heroine of the quest to discover the structure of DNA.

Franklin's work, however, did not take place in a vacuum; she is part of a long list of scientists who preceded her, then followed her, and eventually made the genetic manipulation of living organisms a reality. Before Franklin could do her work on the three-dimensional structure of DNA, however, others had to elucidate its chemical nature. What follows is the story of how they did it.

The Chemical Nature of Genes

We just saw that the theory of the gene, as developed by Mendel, Morgan, and others, could be used to solve many questions pertaining to heredity without really knowing the chemical nature of the genetic material. In the 1930s and 1940s, scientists started to actively investigate the actual chemical substrate of heredity. Mendel had demonstrated that genes were elements of heredity, and Morgan and his collaborators had shown that genes are located on chromosomes. What then are chromosomes made of? An answer to this question would put heredity squarely in the realm of chemistry.

Chromosomes in living cells stain with dyes specific for deoxyribonucleic acid (DNA) and protein. They also contain traces of ribonucleic acid (RNA), a chemical similar to DNA. These are the three classes of very large molecules found in living systems. Since vitalism, the idea of a mysterious life force, had been dismissed convincingly decades earlier by the French scientist Louis Pasteur, it was reasonably assumed that heredity was stored in some as yet unknown way in living cells themselves, somewhere in their chemical makeup. What is more, this type of information, the coding of what a cell or organism was instructed to do during growth and development, could not possibly have been stored in molecules as small as water, sugars, lipids, or other chemicals prevalent in cells. The chemical or informational diversity needed to explain the enormous variety of life forms was simply not present in these simple, albeit ubiquitous, chemicals. Therefore, the attention of scientists shifted to large molecules, called *macromolecules*, like DNA, RNA, and proteins. These could, through their complexity, store genetic information and propagate it to the progeny of cells and organisms through simple division or sexual reproduction. DNA was localized uniquely to chromosomes, making it an excellent candidate as a chemical support for genes. Unfortunately, it was at first thought that plant cells, contrary to animal cells, did not contain DNA. Similarly, some bacterial species were erroneously shown to be devoid of DNA. We know now that flawed techniques were responsible for these faulty conclusions. Nevertheless, because of these studies, DNA was not considered to be a good candidate for the storage of genetic information—it was thought not to be present in all living cells. Moreover, some reports—now known also to be in error—claimed that DNA was not a large molecule at all; it was a small molecule produced by a secondary and unimportant metabolic pathway.

Likewise, even though RNA had been found to be associated with chromosomes, known by then to be the matrix and physical support of inheritable genes, most RNA was found in abundance elsewhere in the cell, in particular in the cytoplasm, the cell sap. Since RNA could not be pinpointed on chromosomes alone, it too was dismissed as a candidate for containing genes. The only choice left was proteins. In the 1930s and 1940s, fractionation techniques that allowed the separation of different classes of molecules showed unquestionably that proteins came in a multitude of forms. The chemical properties of proteins differed enormously among organisms belonging to different species. Certainly, most researchers felt that genes were proteins. A significant problem was that proteins were even more prevalent outside chromosomes than they were inside of them. Nonetheless, the idea stuck until 1944 when the old hypothesis was shattered by a New York-based Rockefeller University research group.

Oswald Avery was the leader of the Rockefeller University team that demonstrated that genes are made of DNA and nothing else. For these experiments, the team did not use plants or *Drosophila* fruit flies; they used the bacterium *Diplococcus pneumoniae* as their model organism. Why use bacteria instead of pea plants or fruit flies, both of which had proved so successful in earlier genetic work? The answer is that another researcher had made a discovery using this bacterium that intrigued the Rockefeller University team. In 1927 a British physician named Frederick Griffith had published strange results of an experiment. He discovered that bacteria can exchange a substance that modifies their hereditary properties, including their ability to cause disease. His research showed that dead *Diplococcus pneumoniae* bacteria, the causative agent of pneumonia, would still infect and kill mice when injected together with a live, harmless variant of the same bacterial species. This variant was a mutant impaired in the production of a compound normally present in the capsule, a mucus-containing structure that surrounds the bacterial cell. The mutant divided perfectly well but did not cause disease. This mutation, a heritable change in a gene, caused the harmless phenotype: Nonvirulence of the bacteria was due to the synthesis of an abnormal capsule. In his injection experiment, Griffith demonstrated that dead, originally virulent bacteria with a normal capsule could restore virulent properties and a normal capsule to the harmless mutant. His conclusion was that the dead bacteria donated some "principle" to the live, harmless bacteria that made them virulent again. What was the "principle" that dead cells transferred to the mutants to restore their virulence? Griffith did not solve the problem, but he had established an experimental system that would enable Avery and his group to tackle the question successfully.

What was Avery's reasoning? He hypothesized that the *D. pneumoniae* system might help him crack the mystery of the genetic material and decide whether it was made of DNA, RNA, or a protein. If DNA was the genetic material, it should have been able—all by itself—to turn the mutant, avirulent bacteria into virulent ones. If not, the conversion should not have taken place. By the early 1940s biochemists had developed techniques that allowed the preparation of biological macromolecules in reasonably pure form. Avery and collaborators prepared a DNA fraction from virulent *D. pneumoniae* and incubated avirulent, mutant cells with this solution. It worked. DNA isolated from virulent cells was able to restore virulence and normal capsule formation in the mutant cells that had incorporated it after acquiring it from the medium. They had identified the agent responsible for hereditary properties; and this agent was DNA. There it was—the answer: Genes were made of DNA.

Today such a discovery would earn its authors instant recognition and a Nobel Prize. However, these rewards were not given to Avery and his coworkers. The scientific community simply was not ready to see genes as DNA rather than proteins, and, despite the fact that Avery had run many controls in his experiments, other scientists claimed that protein contaminants in his DNA preparations were responsible for the results. Refusal to accept Avery's discovery was not just the consequence of scientists' single-mindedness. As I mentioned earlier, proteins, given their great diversity, were thought to be the genetic material. They are composed of strings of twenty different building blocks called *amino acids*, and proteins come in an enormous variety of lengths and composition. DNA was known in 1944 to be a large molecule, composed of phosphate, the sugar deoxyribose, and only four different nitrogenous (nitrogen-rich) bases called adenine, guanine, cytosine, and thymine, which were somehow linked together. Nobody could see how strings made of only four different bases could supersede the diversity afforded by proteins whose strings comprise twenty different building blocks.

The problem, however, was that Avery and collaborators were right and the others were wrong. It took another eight years for Avery to be vindicated. The work of Alfred Hershey and Martha Chase demonstrated that the genetic material of a bacterial virus, its genes, was indeed DNA. Hershey, in contrast to Avery, did receive the Nobel Prize.

An important corollary of the Rockefeller University team's work was this: DNA was able to penetrate cells in spite of its enormous size. In fact, a stretched out, isolated DNA molecule is longer than a bacterial cell. In a way, Avery and his collaborators had performed the first successful genetic engineering experiment in the modern sense. This achievement was not appreciated at the time, not even by the authors of the work. The phenomenon by which inheritable characteristics of cells are modified through DNA uptake was called *transformation* and this terminology is still in use today.

The Structure of DNA

Once the scientific community accepted the idea that genes are made of DNA, the race was on to discover its three-dimensional structure. What was the nature of the DNA molecule? First, it should have such properties as faithful, perfect replication (duplication) to ensure that the descendants of a cell all contain the same genetic information. Earlier biochemists had shown that DNA was a long polymer consisting of very large numbers of deoxyribose molecules linked in a linear fashion by phosphorus-containing chemical bonds called *phosphodiester bonds*. Further, each deoxyribose portion was also chemically linked to one of four possible nitrogenous bases—adenine, guanine, cytosine, and thymine. The exact chemical structures of deoxyribose and of the four bases were also known.

Then, in 1950, the Austrian-born scientist Erwin Chargaff, also of Rockefeller University, discovered an interesting property of DNA. He isolated DNA samples from yeast, cow organs, human cells, and a bacterial pathogen that infects chickens and analyzed them for the proportion of nitrogenous bases they contained. He found two things: First, the amount of adenine (A) in a given DNA sample was always equal to the amount of thymine (T), whereas the amount of guanine (G) was always equal to the amount of cytosine (C). Second, the sum of the amounts of A and

G was equal to the sum of the amounts of C and T. These observations are now known as *Chargaff's rules* and are written

A = T and G = C
and
A + G = C + T

Nobody understood why the DNA bases followed these rules, but this certainly meant that somehow some type of order or regularity must exist in the DNA molecule. These bases were not arranged randomly.

Today we know that the structure of DNA is a double helix in which the bases form centrally located pairs where A faces and interacts with T and G faces and interacts with C through chemical hydrogen bonds. The deoxyribose portions and the phosphate bonds are on the outside of the helix and form DNA's sugar-phosphate backbone. This model accounts perfectly for Chargaff's rule that A is always equal to T and G is equal to C.

The DNA double helix has achieved icon status. It can be found in numerous biotech company logos, is used to decorate T-shirts, and, by and large, symbolizes modern biology. The history of the double helix discovery has been the subject of several books, one of which (*The Double Helix* by James D. Watson)[1] was adapted for the screen (I don't think the movie was a great box office success, but it is well worth watching). The double-helical structure of DNA (published in an article in *Nature* in 1953) is now universally called the Watson-Crick model after its famous discoverers, Jim Watson and Francis Crick. As we saw at the beginning of this chapter, this name unfortunately ignores the fact that crucial X-ray data were obtained by another researcher, Rosalind Franklin. Without Franklin's exquisite work on the diffraction of X-rays by DNA crystals, Watson and Crick simply would not have been able to build their model.[2] Watson, Crick, and Maurice Wilkins (Franklin's supervisor at King's College,

London) shared the 1962 Nobel Prize for this monumental discovery. Rosalind Franklin died prematurely in 1958 without sharing the honor rightfully due her. Fortunately, her considerable merit is now recognized.

What are the biological consequences of a double-helical structure for the genetic material? The short article published in the journal *Nature* by Watson and Crick that reported the discovery of the double helix concludes with this ominous sentence: "It has not escaped our notice that the specific base pairing we have proposed immediately suggests a possible copying mechanism for the genetic material." Indeed, the double-helical model showed how exact duplication of genes is possible: The two strands of the double helix separate and each single strand is used as a template to create two identical copies of the original double-stranded molecule, as seen in Figure 2.1. This mechanism is actually quite complicated and involves numerous protein enzymes (catalysts) to function. The key, however, resides in the properties of base pairs. Every time the enzymes encounter an A (from A, G, C, T), for example, on the template strand, they will insert a T in the growing strand. The opposite template strand will contain a T that originally faced the first A and the enzymes will insert a new A in the second growing strand. In this way, the original A-T pair is fully recreated in the two daughter double helices. This holds true for all the bases, and when DNA is fully replicated, the two daughters, or offspring double helices, have the same base pair sequence and hence carry the same genetic information.

The base "reading" mechanism is extraordinarily precise, but it is not perfect. Sometimes, the enzymes misread a base pair and insert a different base pair in the daughter strands. This is called a *point mutation*, and we now know that mutations are responsible for the existence of different alleles corresponding to a single gene. For example, the white-flowering peas used by Mendel were mutants of the purple ones. The mutant peas produced no flower pigmentation because a gene responsible for purple pig-

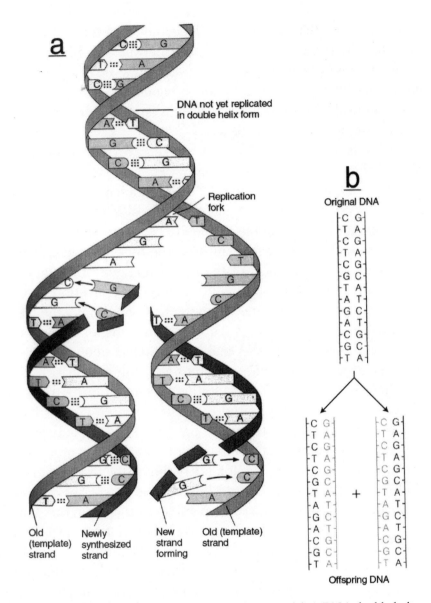

FIGURE 2.1   The Mechanism of DNA Replication. (a) A DNA double helix in the process of being replicated. (b) A flattened representation of the original DNA double helix (top) after full replication (bottom). The two daughter or offspring double helices (the products of replication) have the same base pair sequence and therefore carry the same genetic information (after Brooker, R. J. 1999. *Genetics: Analysis and Principles*. Menlo Park, CA: Benjamin/Cumming).

ment synthesis had mutated and become inactive. This also explains why the F1 of a genetic cross between a white mutant plant and a purple wild-type plant gives purple flowers: The single normal allele (P) provides enough pigment to make the flowers look purple. This is why the P allele is dominant; it imposes its activity on the inactive p allele.

The Watson and Crick discovery marked a turning point in the history of genetics and started what we know today as molecular genetics or molecular biology. The Mendelian gene was no longer a mysterious element without physical foundation; it was made of base-paired, double-stranded DNA that could replicate and be passed on to progeny, most of the time without change, which explained why most of the time genetic information was stably inherited. Once in a while, however, replication errors would lead to the appearance of mutants. For example, the albino allele in humans is one such mutation. Furthermore, even though the variation in DNA makeup, as found among different organisms, came from the permutation of only four bases, it is the sheer length of this molecule that accounts for the tremendous genetic variability found in nature. Simple viruses have DNA containing a few thousand base pairs, whereas bacteria contain millions, and more complicated unicellular organisms contain tens of millions. The totality of human DNA (the human genome) contains more than 3 billion base pairs. The record, however, is held by plants; some plant species contain one hundred times more DNA than humans. This is not because these plants are more complicated or advanced than humans. Rather, it is often the case that most of the DNA in complex multicellular organisms is not genetically meaningful. For example, up to 95 percent of human DNA does not code for genetic information. This still leaves humans equipped with about 30,000 genes, in contrast to 13,600 for the fruit fly, 19,000 for a nematode (a very small worm), and 6,000 for yeast. Therefore, many plants contain a higher percentage of noncoding DNA (often called *junk DNA*) than other living organisms, including humans. The function of the large amount of junk DNA in many plant species is not understood, but we know that this portion of their DNA does not comprise genes.

At this point, we understand that DNA stores the genetic information that defines the nature of a cell; but how does DNA communicate its instructions to the rest of the cell? In other words, DNA is the "instruction manual" that tells a cell what to do. So how is this manual read, understood, and implemented?

The Flow of Genetic Information in Cells

DNA can be thought of as an old-fashioned computer tape. It is a very long ribbon, tightly packed inside cells, where the ones and zeros of binary language are replaced by the four DNA bases (A, G, C, and T). All genetic information—think of it as the programming, or software—of a cell is carried by this tape. The program is then read by the hardware, just as in a computer. This hardware is extremely complicated and can be subdivided into two processing steps: transcription and translation. The flux of genetic information can be diagrammed like so:

First, the genetic information (the genotype) stored in the DNA in the form of long base sequences is transcribed into another macromolecule, RNA, which is very similar to DNA. An exact RNA copy of DNA is made through transcription by enzymes (protein catalysts) called RNA polymerases. RNA contains the sugar ribose (the R in RNA) instead of deoxyribose (the D in DNA) and the nitrogenous base called uracil instead of thymine (plus of course A, G, and C) and is single-stranded rather than double-stranded. What is important is the fact that the genetic messages present in DNA and RNA are identical.

This transcription mechanism may look like a waste of time and energy for living cells. Why make an RNA copy that contains exactly the same information as the original material and in the same form? Why isn't genetic information transferred directly from DNA to proteins? One plausible answer is that more than 3.5 billion years ago, when what was to become cellular life first emerged, the genetic material was not DNA; it *was* RNA. In that origin of life scenario, RNA was the first biological macromolecule to appear from simpler chemicals and to direct the synthesis of proteins and then of DNA itself. In that sense, RNA may be a molecular fossil still inhabiting all of us, all the bacteria and animals that share the planet with us, and all the vegetables we eat.

In the next step, the base sequences now present in RNA form must be deciphered, or decoded, in order to yield the end product of the information transfer process, proteins. Proteins are the engines that drive a cell's activities, such as dividing and producing the specific compounds necessary for survival through metabolism. They perform this task by providing physical support for cellular structures and by acting as catalysts that make possible the many chemical reactions occurring in living cells. Proteins are chains of amino acids linked together by chemical bonds. They differ from one another by their length and amino acid sequence. There are exactly twenty different amino acids used by all cells to make proteins. The question is, When a protein molecule is synthesized, how does the hardware of the cell know which amino acid to insert and where? It was discovered in the 1950s that a mutation in the DNA (the change of a base pair into another base pair) resulted in the positioning of a "wrong" amino acid in a protein molecule. It turns out that the DNA sequence (via its RNA intermediary) determines which amino acid goes where in a protein molecule.

It was soon realized that the genetic code that controls this activity consists of blocks of three contiguous bases called *codons*. Codons determine the kind of amino acid that is incorporated at a

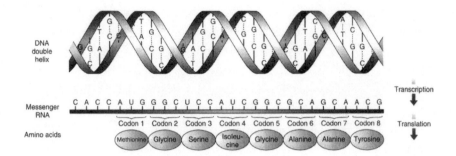

FIGURE 2.2   A Simplified Summary of the Flow of Information from DNA to Protein. The DNA double helix is transcribed into a single-stranded RNA molecule (called a messenger RNA) with a base sequence identical to that of the DNA. Note that a U replaces a T in RNA. Blocks of three contiguous bases in RNA, called codons, are then translated (read) by the translation apparatus of the cell, and the amino acids specified by each codon are linked together to form a protein molecule (after Alcamo, I. E. 1996. *DNA Technology: The Awesome Skill*. Dubuque, IA:Wm. C. Brown Publishers).

given location within a protein. (Francis Crick, of double-helix fame, was a codiscoverer of this mechanism also.) In other words, the bases are read in groups of three by the translation machinery, which itself is composed of RNA molecules working in conjunction with many different proteins. The end result is the synthesis of new proteins whose amino acid sequences are entirely determined by the sequence of codons present in the DNA and its RNA copy, as illustrated in Figure 2.2. The mechanism of translation actually *does* translate RNA language (bases arranged in codons, exactly as in DNA) into protein language (amino acids). Newly formed proteins then go on to perform their functions as metabolizers, for example, or builders of cell structure. These functions constitute what is called the *phenotype*, in short, what cells look like, what they do, and how they respond to stimuli. The genotype (the DNA) in the form of a base sequence determines the phenotype, itself represented by amino acid sequences in proteins.

Since the genetic code is universal (with very minor exceptions), genes from any source contain codons that can be understood by any organism or cell that receives them. For example, a string of codons contained in a bacterial gene will be correctly interpreted by a plant's translational apparatus and vice versa. Thus, it should be possible to engineer plants with bacterial genes. We will see in Chapters 3 and 4 that, indeed, this feat can be done but that in reality things are a little more complicated than simple codon deciphering.

The grand conclusion of this story is that the genetic manipulation of an organism can be achieved theoretically provided that (1) cells in that organism can be coaxed to incorporate externally supplied DNA (that is, they can be *transformed* by DNA uptake), (2) the externally supplied DNA is enriched in genes corresponding to the traits the experimenter wants to add, and (3) the foreign genes are properly transcribed and translated in the new environment. In a nutshell, this is how biotechnology works.

We have seen in the first two chapters that biotechnology would have been impossible without an understanding of inheritance (Mendel's and Morgan's work), the discovery that cells can take up DNA and that DNA is the genetic material (Avery's work), and an understanding of the flow of genetic information (Watson, Crick, and many others).

It took about 100 years to unravel the mysteries of the gene, from its discovery by Mendel to the deciphering of its code, the blueprint of life. The science of genetics was now poised to take the next step, gene cloning and manipulation. In other words, genetics before gene cloning had been largely a "contemplative" science because geneticists had little or no controlled access to the fabric itself of the genetic material. They could study it, but they could not change it easily in a specific manner. The recombinant DNA revolution changed all that; genes would become a material that one could modify and rearrange almost at will.

Rosalind Franklin, the scientist who did so much to unravel the three-dimensional structure of DNA, died prematurely of cancer before her contribution was recognized. Today, with the hindsight of several decades, Franklin's achievements are finally acknowledged. In later editions of his book *The Double Helix*, Jim Watson, who offered less-than-flattering details of Franklin's personality in the first edition, finally recognized her critical role in the elucidation of the DNA double helix. Her legacy is now clear: Nearly all modern genetics textbooks show her X-ray picture of the DNA molecule and often, although not often enough, her portrait.

3

Genes in the Test Tube: The Recombinant DNA Revolution

IN 1980 MY COLLEAGUE at Washington State University, geneticist Andy Kleinhofs, wrote for all to see on his laboratory chalkboard what he called the Eleventh Commandment: "Thou shalt clone and sequence." Kleinhofs is close to seven feet in height and tends to dominate the debates in which he takes part, so we took this Eleventh Commandment seriously. He had summarized for the benefit of his graduate students and others what the foundation of the genetic motto of the 1980s turned out to be: the cloning of genes and the determination of their base sequence.

Kleinhofs was interested in a barley gene coding for the enzyme nitrate reductase, an important component of the cell machinery that assimilates the fertilizer ammonium nitrate. One of his distant goals was to isolate that gene, manipulate it for better efficiency, and then reintroduce it into plants. But first, the gene had to be cloned and sequenced, as per the directives of the Eleventh Commandment. In order to achieve this goal, Kleinhofs used a set of techniques called *recombinant DNA technology* that had been developed a few years earlier. This chapter explains how the technology works and how researchers today can clone practically any gene from any organism.

41

Recombinant DNA technology defines our ability to isolate single genes or groups of genes and clone them. Cloning[1] involves separating certain genes from all other genes in an organism and multiplying them in living cells or in the test tube. This process is not trivial. Even the simplest bacteria contain thousands of genes coding for thousands of different features like metabolism, virulence, shape, and adaptation to the environment. It is impossible to study and sort out all the properties of these genes "in bulk." In order to understand what single genes do, they must be isolated from all other genes and studied individually. DNA recombination and gene cloning allow just that. They also allow the multiplication of these isolated genes that facilitates their introduction into unrelated hosts. The results stemming from these manipulations are genetically engineered organisms, often called *genetically modified organisms* or *GMOs*. GMOs could not exist without the technique of gene cloning. It is important to know *how* genes are cloned in order to understand the nature and origin of GMOs. The concepts developed in this chapter are essential to establish a good grasp of what biotechnology is.

As we saw in Chapter 2, Avery and his coworkers successfully transformed bacteria with purified DNA in 1944, well before the structure of DNA had been elucidated. Their positive results were due, in part, to a good dose of serendipity. Indeed, their test bacterial organism, *D. pneumoniae*, is now known to be one of very, very few that spontaneously absorb DNA when it is added to growth medium in test tubes. Had they used one of the other bacterial strains then commonly used in research laboratories, they would have ended up with no results at all. There are virtually no cells or organisms that take up DNA spontaneously; they must be forced to do so, and we will see later how this work is achieved.

Nevertheless, Avery and his collaborators discovered DNA-mediated transformation, a cornerstone of biotechnology, and this discovery was a giant first step that inspired others to find methods to introduce DNA into various types of cells. However, incubating living cells with *all* of the DNA isolated from other cells, as Avery et al. did, is a very inefficient way to achieve the genetic modification of organisms. This technique is no longer used today and has been replaced by a method involving highly concentrated parts of the DNA, specific genes, coding for the desired traits. It is called gene cloning.

This chapter describes the basics of gene manipulation and in particular, the fundamentals of gene cloning, gene identification, and gene structure. The principles and methodologies described here are not specific to plants; rather, they apply to gene manipulation in any type of organism. It goes without saying that cloning techniques are used extensively in the genetic engineering of plants, and this principle must be grasped in order to understand plant biotechnology.

Cloning

Biotechnology would be impossible without gene cloning. This technique allows the isolation and multiplication of single genes, starting from the total DNA extracted from an organism. *Prokaryotes*—cells without a nucleus to contain their DNA, such as bacteria—contain roughly 2,000 to 4,000 genes. *Eukaryotes*—organisms whose cells do possess a nucleus, such as the one-celled yeast, oak trees, and humans—contain more genes, tens of thousands in the latter two examples. One can immediately see the challenge of fishing a single, specific gene out of that mass of DNA. Yet, this task can be accomplished today thanks to a series of discoveries made in the late 1960s and early 1970s. These discoveries were not what is called mission-oriented. In other words, it was not the intention of the investigators to develop biotechnology; they were simply trying to decipher the basic mechanisms of life. Cloning became possible thanks to fundamental work first done in the field of bacterial genetics. The whole enterprise started with geneticists who were interested in the bacterium *E. coli* (short for *Escherichia coli*) and explored its inner workings.

E. coli became the workhorse of bacterial genetics in the 1940s. In rich growth medium, it can divide in as few as twenty minutes, producing over 1 billion cells per ¼ teaspoon (about one milliliter) of broth in just a few hours. It is also the most common inhabitant of the human gut and is responsible for the foul odor of feces. However, laboratories where *E. coli* is used should not be thought of as repugnant dens where masochistic researchers carrying fiendish experiments dwell: *E. coli* grown in the presence of oxygen (that is, in the presence of breathable air) does not produce the compounds responsible for the strong smell of excrement.

To continue, several *E. coli* strains isolated in France in the 1940s were given designations starting with the letter M (such as ML 3), from the Latin "merda," meaning "merde" in French and its vernacular complement in English. The "L" in "ML" reputedly originated from the name of André Lwoff, the French Nobel laureate. These early bacterial geneticists were evidently dedicated, body and soul, to their work. Given its high rate of division and absence of pathogenic (at least in the case of lab strains) and other noxious effects, *E. coli* became the model for the study of bacterial genetics. It could also be easily isolated from human waste if necessary. Pathogenic *E. coli* strains do exist in nature and occasionally infect humans, with disastrous consequences. Scientists *never* use these dangerous strains to perform gene cloning experiments. We will see later that scientists have been aware of the potential dangers of gene cloning since 1975 and have taken drastic measures to minimize its risks.

The first element necessary to conduct cloning experiments is something called *restriction enzymes*, also called *restriction endonucleases*, that were discovered by studying bacterial viruses. Bacteria, just like all other types of organisms, can be infected and killed by viruses. Viruses that kill bacteria are called *bacteriophages* (from the Greek for "bacteria-eating"). In the 1950s, scientists discovered that some *E. coli* strains that were isolated from different sources were immune to bacteriophages; the viruses simply would not reproduce in these cells. About ten years later the reason was discovered: The immune *E. coli* strains possess specialized enzymes called *restriction enzymes* that cut the DNA of the invading bacteriophages and rendered it inoperative. These enzymes can be thought of as a defense mechanism that some bacteria developed to resist viral infection. We know today that restriction enzymes are extremely common in the bacterial world.

It turned out that these restriction enzymes do not cut DNA randomly. They recognize specific parts of the DNA double helix, that is, specific base sequences, and cleave the DNA at these locations. This means, for example, that a piece of DNA containing a single recognition site will generate two fragments after cleavage. A piece of DNA containing two such sites will generate three pieces after cutting, and so on. Cutting DNA from any source with restriction enzymes produces a collection of fragments with different lengths, depending on the DNA sequence and the number of restriction sites present in the sequence.

Dozens of restriction enzymes from many bacterial sources are presently available commercially and are used by scientists to generate clonable DNA fragments, that is, pieces of DNA that are not impractically long. Restriction enzymes are used by biotechnologists to cut long DNA molecules into shorter fragments that contain a limited number of genes. This step makes the process of gene cloning and identification much easier because dealing with chromosome-length pieces of DNA containing thousands of genes is cumbersome if not impossible.

At roughly the same time, other investigators discovered genetic elements in bacteria that harbored genes not linked to genes present on the bacterial chromosome. The bacterial chromosome is a long, single piece of circular double helical DNA. In a typical bacterial cell, up to 4,000 genes are contained within this single DNA molecule. However, some bacterial genes behaved in experiments as if they were not linked to this chromosome. Their behavior indicated that they were located on a separate piece of DNA (a "minichromosome," if you will) independent from the main chromosome. In this sense, these bacterial cells contain one large chromosome and one small one. And indeed, electron microscope studies demonstrated that many bacterial species, including *E. coli*, possess such minichromosomes, now called *plasmids*. Thus, these plasmids are physically distinct from the main chromosome. Plasmids often code for accessory, or nonessential, functions that can be beneficial to bacterial cells harboring them. They are, however, frequently dispensable, as most of them are not necessary for survival. Examples of genes carried by plasmids include those for antibiotic resistance (an annoying property that makes the bacteria harboring them insensitive to antibiotics), heavy metal detoxification, and degradation of synthetic organic compounds. They are also often short, with lengths as brief as a few thousand base pairs (as opposed to millions of base pairs for the main chromosome), and possess the ability to replicate independently and often profusely in the bacterial cell. Like the bacterial chromosome, plasmids are pieces of circular, double-stranded DNA.

Gene cloning basically means multiplying, isolating, and purifying a gene or genes, usually using *E. coli* cells. In 1972 Stanley Cohen of Stanford University and Herbert Boyer of the University of California realized that they could combine plasmids and restriction enzymes to clone any piece of DNA from any source in *E. coli*. For this procedure, they used plasmid DNA, opened it up with a restriction enzyme, combined—spliced it—with the DNA to be cloned (in a sense, the DNA rides piggyback on the plasmid), and introduced the new molecules into *E. coli* for replication and multiplication. This technique worked; and it did so for the following reasons:

1. The plasmid, used as a vector (vehicle, if you will) to carry pieces of DNA to be cloned in *E. coli*, must have a single recognition

site for a restriction enzyme. When this is the case, a circular plasmid will be converted into a linear DNA molecule with the same length as the circular one (see Figure 3.1).

2. The restriction enzyme used to cut the plasmid should be the same one used to cut the foreign DNA. In this way, the plasmid DNA and the foreign DNA will have the same sequences at their ends, and these ends can interact to form a hybrid molecule when the cut plasmid and the cut foreign DNA are mixed. The new molecule will consist of the plasmid itself now attached to a piece of foreign DNA. An enzyme called *ligase* should then be used to suture the DNA sugar-phosphate backbone and recreate a circular, now hybrid, molecule called a recombinant DNA molecule.

3. Recombinant DNA molecules must then be introduced into *E. coli* cells by transformation, where they will replicate and multiply themselves.

All the needed ingredients and techniques existed when Cohen and Boyer started their experiments. They used a plasmid called pSC101 that confers resistance to the antibiotic tetracycline to cells hosting it.[2] This plasmid contains a single cutting site for the restriction enzyme *Eco*RI (so named because it was isolated from *E. coli* strain RI). Cohen and Boyer then prepared and made the plasmid linear with the enzyme. This preparation was mixed with another *Eco*RI-cut plasmid called RSF1010. The second plasmid imparts resistance to another antibiotic, streptomycin. The mixture of cut pSC101 and RSF1010 was incubated with ligase to make recombinant DNA molecules and presented to *E. coli* cells under a special set of conditions that triggered DNA uptake.

Normally, *E. coli* cells do not take up DNA from the medium, unlike *D. pneumoniae* as used by Avery and his coworkers. Fortunately, other investigators had discovered two years earlier a method that allows *E. coli* to become permeable to DNA. This method involved incubating *E. coli* cells in the presence of calcium chloride at a heat-shock temperature of 41°C. For reasons still not understood, *E. coli* picks up any kind of DNA under these conditions. Cohen and Boyer used the calcium chloride–heat shock technique to introduce the recombinant pSC101/RSF1010 plasmid into their *E. coli* cells. The whole cloning process is diagrammed in Figure 3.1, which shows the original experiment. Cohen and Boyer had succeeded in joining two DNA molecules together, in the test tube, in order to generate a new, single hybrid molecule that could not have been produced in nature. Their transformed *E. coli* cells had become resistant to *both* tetracycline *and* streptomycin because both resistance genes had become linked on a single DNA molecule.

This experiment marked the birth of biotechnology. Scientists had discovered a means to attach a piece of DNA to a plasmid, called the vector, which, when transformed into *E. coli*, replicated perfectly normally, as if the other piece of DNA were not even there. The Cohen team demonstrated quickly thereafter that DNA isolated from the giant clawed toad (*Xenopus laevis*) could be cloned and propagated similarly in *E. coli*. In other words, a piece of amphibian DNA was now an integral part of a bacterial genome. There is no need to focus on the use of toad DNA, as opposed to any other DNA, for this cloning experiment because the type of DNA is not important to the result. The toad DNA just happened to be available in the lab at the time of the experiment. It seemed likely to the researchers that if toad DNA could be cloned in *E. coli*, then any DNA from any source could also be used. Indeed, this assumption turned out to be true: DNA from all possible sources, including human, has been cloned in *E. coli*.

In summary, the Cohen-Boyer experiment showed that two pieces of completely unrelated DNA could be linked in the test tube and propagated in *E. coli* cells. Since these recombinant molecules multiplied abundantly in *E. coli*, it had become possible to purify and study them at the DNA sequence level, giving precious indications regarding the nature of the cloned foreign genes.[3]

It should be understood that *E. coli* cells containing some toad DNA do not look like bacterial/toad hybrids. On the contrary, *E.*

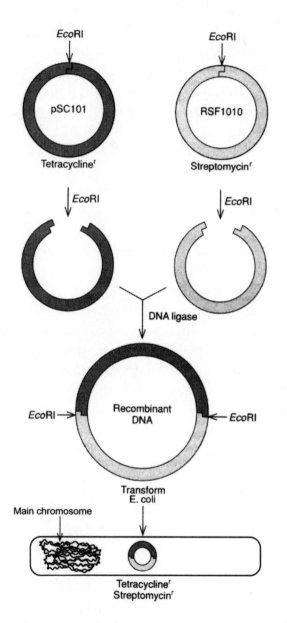

FIGURE 3.1   The First Boyer/Cohen DNA Cloning Experiment. This was a "proof-of-principle" experiment aimed at demonstrating that the idea of linking two DNA molecules in the test tube and multiplying these recombinant molecules in E. coli would actually work (after Weaver, R. F., and P. W. Hedrick. 1991. *Basic Genetics*. Dubuque, IA: Wm. C. Brown Publishers).

coli cells containing plant, human, fungal, or any other DNA look like perfectly ordinary *E. coli*. There are two reasons for this appearance. First, even today, it is only possible to clone relatively short stretches of foreign DNA in *E. coli* or in any other organism. This means that only a limited number of foreign genes (up to a few hundred) can be incorporated into *E. coli*. Second, eukaryotic genes, such as those originating from a toad, a plant, or a human being, are not expressed in bacteria—they remain silent. In order for foreign genes to be expressed, they must be further manipulated. We will explore these processes later in the chapter.

The question arises, If *E. coli* could be engineered with foreign genes, why not immediately apply this technology to plants? One conceivably could clone foreign genes in *E. coli*, multiply them there, purify them, and reclone them in a vector suitable for plants. These recombinant DNA molecules could then be transformed into plants where they would be expressed. In 1972 when Cohen and Boyer made their discovery, there were two major obstacles standing in the way of such experiments. First, bacterial plasmids, as vectors, were not expected to replicate in plant cells (we now know they don't) because of the vast evolutionary distance separating bacteria and plants. Nobody knew what kind of vector to use. Second, nobody knew how to introduce DNA into plant cells, in other words, how to transform them. The solutions to these two problems came just a few years after the Cohen-Boyer breakthrough and marked the birth of plant genetic engineering. This technology will be described in Chapter 4.

Another Technique

An additional and very important molecular technique worth discussing is the ability to sequence the base pairs present in DNA. Since the genetic code relating amino acids to DNA bases is known, the base sequence of a gene automatically provides the amino acid sequence of the protein for which it codes. Given these sequences, it is possible to make predictions regarding the function and properties of a gene and its protein product. This is an important consideration in the genetic engineering of plants for practical purposes. Further, knowledge of the sequences of all the genes present in an organism ultimately can tell us what makes the organism "tick" and can lead to the cloning of specific genes.

Nobel laureates Frederick Sanger of England and Walter Gilbert of Harvard University developed DNA sequencing techniques in the late 1970s. These techniques are now routinely used in many laboratories and have been massively applied to the resolution of the full base sequence of human DNA (3.15 billion base pairs) under the aegis of the Human Genome Project. The latter is a multinational effort to map the DNA sequence of all human genes; its first phase, the sequencing of human DNA, was completed in June 2000. The detailed identification of all human genes present in this sequence will take just a few more years. To date, the full base sequences of about two dozen bacterial genomes are known, as are the sequences of yeast, the fruit fly *Drosophila melanogaster*, and the nematode *Caenorhabditis elegans* (a very small worm).[4]

The full sequence of the first plant genome, that of the plant *Arabidopsis thaliana* (genome size = 130 million base pairs containing 25,000 genes) was finished at the end of the year 2000. Among plants, this genome is very small and, hence, was relatively easy to sequence. Moreover, *Arabidopsis* is a genuine complex plant with leaves, roots, flowers, and seeds and is very often used as a laboratory model for all plants. Other than that, its only other notable attribute is its use as a summer salad ingredient in some European countries. The complete sequencing of the *Arabidopsis* genome will considerably accelerate the sequencing and understanding of crop plant genomes. This progression, in turn, will allow the identification and cloning of genes of agronomical interest, such as those coding for yield, disease resistance, and frost tolerance. In addition, a rough draft of the rice genome recently has been obtained. These discoveries will great-ly facilitate the cloning of genes that scientists want to introduce into a variety of crop plants.

DNA sequencing is now fully automated. Robots mix the DNA to be sequenced with the necessary chemicals used in sequencing experiments, and the sequence of bases is read by another machine called a *gel reader* that provides a graphical output of the results. Laboratories equipped with several such machines can sequence tens of thousands of bases or more per day. Ten years ago researchers considered themselves happy when they could sequence just 500 bases in one day.

Identifying Cloned Genes

We have described how it is possible to clone DNA in *E. coli*. However, cloning pieces of DNA, albeit a major achievement, does not tell scientists exactly which genes are present in the frag-ments of DNA they have cloned. In other words, having cloned a piece of DNA from, say, a plant, one would like to know what gene or genes are present in this DNA fragment and what their biological functions are. Again, it should be remembered that each recombinant vector molecule propagated in an *E. coli* cell contains only some foreign genes, perhaps a few dozens, whereas it requires tens of thousands of genes to constitute a full plant genome. Assigning a biological function to a piece of DNA is a critical issue.

Over several years, sophisticated techniques have been devel-oped to tackle the problem of determining which genes are pres-ent in a cloned fragment of DNA. Today the favored approach is to sequence the piece of cloned DNA and compare its sequence to that of genes with a known biological function. Public electron-ic databases store hundreds of thousands of known gene base sequences in addition to computer software that allows quick alignment and comparison of an unknown piece of sequence with the information present in those databases. This software tells the scientists the percentage of similarity (homology) that exists between a known gene and the unknown DNA. A possible biological function can then be assigned to the newly cloned DNA.

Now, what happens if a piece of carrot DNA is sequenced, run through the computer, and turns out to be homologous to a known hamster gene? Can one have any confidence that such a result that shows similarity of biological function of these genes represents reality? In many cases, the answer is yes. It has been repeatedly demonstrated that a large number of genes present in mammals, for example, have very similar counterparts in plants, and vice versa. This finding should not surprise us; as we understand the tree of life today, multicellular eukaryotes have evolved from single-celled ones, and these have in turn evolved from prokaryotes. Indeed, quite a number of prokaryotic genes also have recognizable homologues in eukaryotes, including humans. Obviously, various species also have specific genes that make them unique.

The function of these specific genes is more difficult to determine. Database comparisons show no "hits" and thus no putative function assignment. How then does one define biological function? One can use a technique called *gene replacement* or *gene knock-out*, by which one replaces the good gene naturally present in an organism with a bad version generated in vitro (in the test tube). Indeed, several techniques allow the production in vitro of mutant versions of a cloned piece of DNA by changing one or several of its base pairs. When these mutant versions are introduced into a normal organism, they replace the normal genes and sometimes produce a distinct, abnormal, phenotype, for example, the arrest of organ development or inability to metabolize a certain compound. When such an anomaly appears, researchers can attribute a loss of biological function to the mutant gene and, hence, decide what the function of the normal gene was.

This procedure has been used in mice bred to contain the gene responsible for the human genetic disease cystic fibrosis. In these experiments, the equivalent mouse gene was mutated in vitro and used to replace the good gene in normal mice. These gene-replaced mice developed symptoms associated with cystic fibrosis and are now used as a model system to study and perhaps find a treatment for the human disease. Conversely, gene replacement can be used to replace a bad gene with a good version. The feasibility of this technique has also been demonstrated in mice, where animals harboring mutated hemoglobin genes (and suffering from anemia) were treated with normal human hemoglobin genes. The mice returned to a normal health. The latter approach is often referred to as *gene therapy* and is now used in humans on an experimental basis.

Unfortunately, gene replacement works imperfectly in plants, making gene function determination much more challenging. This situation is not because plants are more complicated; the problem is (and always has been) inadequate funding for the plant sciences. Increased funding would provide research into better techniques that would improve gene replacement in plants. Another problem is that relatively few researchers are interested in plant biology. However, there do exist other techniques (though they are less efficient) to tackle the problem of gene identification in plants.

The Fine Structure of Genes

Researchers realized quickly that prokaryotic genes cloned in *E. coli* would generally be active and produce the proteins for which they coded. However, eukaryotic genes would not. This finding was puzzling because it was known that the genetic code was universal and, hence, the *E. coli* translation apparatus should have been able to read and understand genes isolated from eukaryotes. Similarly, animal and plant cells, for example, transformed with bacterial genes would not express these genes. What were the differences between prokaryotic and eukaryotic genes? The sequencing of DNA from prokaryotes and eukaryotes provided the answers.

All genes, eukaryotic and prokaryotic, contain three major elements: the promoter, the coding sequence, and the terminator. The promoter is the region of the gene recognized by RNA polymerase, the enzyme that produces an RNA copy of the DNA. The promoter sequence is located just before the coding sequence. Without a promoter, RNA polymerase does not know that a given piece of DNA should be transcribed, and it simply will not bind to the DNA. A gene that is not first transcribed cannot have its corresponding RNA translated, since this RNA is never produced. It turns out that prokaryotic and eukaryotic gene promoters are quite different in the organization of their base sequence. Prokaryotic RNA polymerase does not recognize eukaryotic promoters, and eukaryotic RNA polymerase does not recognize prokaryotic promoters. This inability is the reason why bacterial genes, without further engineering, are not expressed in animal and plant cells, and vice versa.

However, human genes *can* be expressed in bacteria (such as the insulin gene), and bacterial genes *can* be expressed in plants. How is this done? Very simply, the promoters of cloned genes are cut away by restriction enzymes and replaced with appropriate bacterial, animal, or plant promoters, all using the recombinant DNA technology. Genes containing the promoter from one organism and the rest from another one are called *chimeric*, after the mythical beast *Chimera*, who was composed of incongruous parts.

The coding sequence of a gene is located on the DNA immediately after the promoter region. This sequence is the one that contains a string of three-base codons, or three contiguous bases in the RNA copies of the gene, that will end up being translated into amino acid sequences, the proteins. Since the genetic code is universal, once an RNA molecule corresponding to a DNA gene is produced, it will be translated, regardless of its origin and new host. This holds true for bacterial genes introduced into other bacteria and works equally well for bacterial genes introduced into animal and plant cells. However, even when equipped with a bacterial promoter, most animal and plant genes are still not expressed in bacteria. This result should not have been a problem for scientific researchers because, given the action of the engineering of a bacterial promoter in front of the eukaryotic coding sequence, an RNA copy of the gene should have been produced. And indeed it was.

The problem here was not transcription, it was *translation*. The vast majority of eukaryotic genes have strange coding sequences: They contain many apparently useless stretches of base pairs. These sequences are called *introns* and nobody really knows their origin. The meaningful (in terms of sequences coding for amino acids) portions of eukaryotic genes are called *exons*. A typical eukaryotic gene can be seen as a number of exons separated by introns. Almost all bacterial genes, on the other hand, are devoid of introns, and their coding sequences, in the form of RNA, can readily be deciphered by the eukaryotic translation machinery. Bacterial genes equipped with relevant promoters are usually expressed normally in a eukaryotic host as in a plant. The reverse is not true because prokaryotes are incapable of processing RNA molecules containing introns.

How do eukaryotic cells deal with RNA molecules in which exons and introns are interspersed? All eukaryotic cells possess systems that clip out introns from RNA and splice the exons together. This process leaves only meaningful base sequences in the RNA transcripts that can then be translated into proteins. Therefore, in order to express eukaryotic genes in bacteria, these genes must be engineered for intron removal. Also, even though all eukaryotes possess the intron-removal machinery, it is not necessarily true that animal introns, for example, will be correctly recognized and removed by plant cell splicing functions. It is often desirable to remove introns from animal genes before introducing them into plants. Thus, engineering bacteria and eukaryotes involves much more than simply cloning a gene; this gene must be tailored with the proper promoter, and intronic sequences should either be removed or their accurate processing in the new host should be verified.

Finally, the coding sequences of genes are terminated by base sequences appropriately called *terminators*. Their function is to tell RNA polymerase to stop transcribing a DNA gene and release its RNA copy. Here also, terminator sequences differ between bacteria, plants, and animals. Consequently, a foreign gene must be equipped with a relevant terminator to be correctly expressed (that is, to make the right protein) in a foreign environment. This further tailoring can also be done by using the recombinant DNA technology.

To use the plant example, if one wishes to express a bacterial gene there, one must replace the bacterial promoter and terminator regions with the corresponding plant sequences. If an animal gene is to be expressed, in addition to promoter and terminator replacement, one must also consider the existence of introns and probably eliminate them before gene transfer.[5]

So far, commercially available genetically modified crop plants have mostly been engineered with bacterial genes. The coding sequences of these bacterial genes were necessarily cloned between plant-expressible promoter and terminator sequences. However, since bacterial genes do not contain introns, these did not have to be removed before engineering. An actual example of a bacterial gene equipped with promoter and terminator sequences appropriate for expression in plants is given in Figure 3.2. The chimeric bacterial *bar* gene is cloned in an *E. coli* vector called pUC 118 that is widely used to perform recombinant DNA manipulations. The *bar* gene determines resistance to a herbicide and is preceded by a plant virus promoter (p35S) that is strongly expressed in plant cells. The terminator (ocs3') is from a gene also designed to work in plant cells. This chimeric *bar* gene has been used to produce plants resistant to a herbicide called Liberty® that will be discussed later.

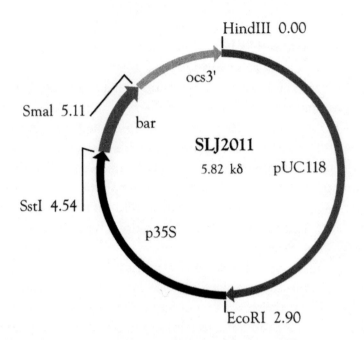

FIGURE 3.2   Chimeric Plasmid SLJ2011 Containing the Plant-expressible, Herbicide-resistance Bacterial *bar* Gene (thick arrow). p35S is the viral promoter and ocs3' is the terminator. The cloning vector is pUC118.

Safety

It soon came to the attention of molecular biologists that the new recombinant DNA technology was not without potential dangers. One could easily conceive that normally innocuous bacteria could be turned into dangerous superbugs once their genomes had been tampered with. After all, evolution had fine-tuned the relationships between bacteria and their ecological niches for billions of years, whereas scientists could now potentially change all that in a matter of minutes. Further, *E. coli*, in which all this cloning was done, is a normal and abundant guest of the human intestinal tract. Could it be that engineered *E. coli* cells would be able to invade humans and colonize their guts, thereby possibly replacing the resident microbial flora? Worse, could it be that foreign genes cloned in these potential invaders would be able to transfer themselves to human intestinal cells? It would be catastrophic if *E. coli* hosted cloned cancer genes, for example.

In 1975 an international group of concerned expert scientists met at Asilomar, a conference center located in beautiful Pacific Grove in northern California, to discuss these possibilities. This meeting became known as the Asilomar Conference, and this was where safety recommendations regarding recombinant DNA experiments were first crafted. These recommendations soon became strict guidelines adopted and implemented by federal funding agencies such as the National Institutes of Health, the U.S. Department of Agriculture, and the National Science Foundation. These guidelines specified which experiments were banned (increasing the virulence of pathogens, for example) and what degree of physical and biological containment was necessary for allowed experiments. Physical containment refers to the type of laboratory facilities necessary to prevent spread of recombinant bacteria, ranging from regular laboratory rooms to negative pressure facilities that prevent escape of laboratory air, depending on the risk involved. Biological containment refers to the actual ability of recombinant bacteria to survive outside of a laboratory environment. As a result of these guidelines, several disabled *E. coli* strains that will not survive outside the laboratory were generated and used. Noting the year of its creation (the bicentennial), one of them was patriotically named X1776. Having used this *E. coli* strain and some of its derivatives myself, I can attest to the great difficulty of keeping them alive, even with all the tender loving care of a laboratory environment. One researcher went so far as to drink a suspension of these crippled *E. coli* cells to demonstrate that they could not survive in the human gut. It is comforting to know that in a span of twenty-five years, there has never been a single reported case of recombinant bacteria having escaped the laboratory and having caused harm. Therefore, one must consider that wisely applied recombinant techniques are safe, at least in the case of *E. coli* and other nonpathogenic bacterial species. This confidence does not mean that vigilance should be relaxed, however.

The year 1972 thus marked the beginning of the biotechnology era. Molecular genetics had given birth to an applied science, and the world would never be the same. Scientists had learned how to manipulate genes and produce recombinant DNA-containing bacteria that could never be produced in nature. There now exist in the world thousands upon thousands of strains of *E. coli* harboring thousands of genes from hundreds of organisms. The vast majority of these strains can be thought of as representing "libraries" of cloned genes and are not used for industrial or medical purposes or for any purposes other than basic research. However, three recombinant strains should be well known to the public: One is used to produce human insulin (which can also be produced by genetically engineered yeast); the second produces human growth hormone; and the third makes plasminogen activator, a substance given to stroke victims. All three products are widely used in medicine. There is no doubt that, seen from that angle, the genetic manipulation of these bacteria was greatly beneficial to certain segments of humanity. This point should be kept in mind when weighing the risks and benefits of genetic technology. Not all of it is negative, as purported by some. Furthermore, as far as I know, insulin- and growth hormone-dependent patients have never opposed the technology—the genetic engineering of bacteria—that permitted the amelioration of their medical conditions.

Animals are also being genetically engineered for medical purposes. One example is the engineering of pigs as organ donors for humans. Pigs and humans are physiologically closely related (all jokes aside), and so pigs could provide replacement organs when human counterparts are not available. The barrier, however, is tissue rejection that results because humans and pigs have different immunological characteristics. Attempts are now being made to engineer pigs in such a way that humans would no longer recognize potential tissue and graft transplants as foreign and would no longer reject them. Another example is the engineering of ewes and goats for the production in their milk of proteins of medical interest, such as blood clotting factors. These experiments have been successful, but presently there is no program that aims at producing these proteins on a large scale. Finally, as far as I know, no one is thinking of genetically engineering animals for the purpose of procuring products that could be used in human nutrition.

Today, we have the ability to genetically engineer practically any life form, using variations on the techniques described here. Scientists today are able to shuttle genes among plants and bacteria, bacteria and animals, animals and plants, bacteria and plants, and so forth. Basically, biological barriers have been abolished, and humans can do what nature never could. This raises many ethical—as well as scientific—questions. We will see later that plant genetic engineering has not met with the same success with the public as that encountered by bacteria engineered for medical purposes.

As for Andy Kleinhofs, he successfully cloned and sequenced his nitrate reductase gene. However, he could not introduce it at that time into other plants, as these techniques had not yet been fully developed. Next, let us explore how these techniques work and how they were invented.

Difficult Beginnings
and Great Triumphs:
The Transformation of Plants

SCIENCE SOMETIMES WORKS in mysterious ways. A researcher can spend years working on a certain topic without attracting the interest and attention of his colleagues. Sometimes, however, a research topic seems to reach some degree of maturity and at that point, stiff competition replaces obscurity. This happened to me in 1979 at an international meeting of plant geneticists held in the town of Szeged, Hungary. At that time, not a single plant had been genetically engineered. In fact, a significant portion of the meeting was devoted to this question: How could one introduce foreign genes into plants?

I had come up with the idea that DNA could be trapped into microscopic fat droplets (called liposomes) and these droplets fused with plant cells. In this way, DNA would be internalized by the cells and its genes perhaps expressed. Experiments supported the view that DNA was indeed incorporated by the plant cells, and I gave a lecture to that effect at the Szeged meeting, convinced that my approach was quite unique. To my stupefaction, the next speaker, Ken Giles of Worcester Polytechnic Institute, Massachusetts, described the very same technique! Giles and I had been working in isolation, and neither one had any inkling of the other's activities. What this meant, however, was that we both were on to something new and perhaps exciting, in contrast to what had been happening so far in the fledgling field of plant genetic manipulation.

We now retrace our steps back in time, but just to 1968. Gene cloning was unheard of, DNA could not yet be sequenced, and E. coli was still not transformable by naked DNA. Yet some researchers had already tackled the problem of genetically modifying plants with DNA extracted from bacteria and from plants themselves. The effect of their results on scientific progress would prove to be both damaging and constructive, all at the same time. Indeed, for years, the field of plant transformation was flooded with irreproducible results that caused many scientists to abandon hope that there was any merit in pursuing such a line of enquiry. Nonetheless, a critical analysis of these flawed reports, in addition to sound investigations, led finally to success.

In many ways, the recorded history of science is an oversimplification of reality. Books on this topic very often give a linear account of the progress of ideas, from simple observations to grand theories, as if the former always led smoothly and directly to the latter. Much of that account is due to the fact that many histories of science deal with events that took place centuries ago, that is, they are describing events that were filtered through the fine sieve of time. False starts and wrong interpretations are seldom mentioned, perhaps because they are not remembered or were never recorded. This is not the case with plant genetic engineering since these events are so recent. The first section of this chapter will give a brief history of the early results obtained in the field of plant transgenesis, the modification of the genetic characteristics of organisms via transfer of externally supplied DNA.

By definition, transgenic plants are plants containing one or several foreign genes introduced by transformation. We discussed previously that the genetic transformation of E. coli, the workhorse of bacterial genetics, was achieved only in 1970. A few other bacterial genera had been transformed earlier, but their DNA uptake mechanism was so special that it did not seem applicable to plants. Obviously, plants are much more complicated than simple *E. coli* or other bacteria. How then, is it possible that plants could be transformed with DNA before the simpler bacterial system was well understood? The answer is straightforward: Plant transformation was in fact *not* discovered in the late 1960s and 1970s. All of the early results in the field of plant transgenesis proved to be wrong. It was only in 1983, about ten years after the events described next, that plant transformation was demonstrated and universally accepted. This great success will also be described.

False Starts

Inspired by the 1944 discovery by Avery and his colleagues of the uptake and genetic expression of externally supplied DNA in the bacterium *D. pneumoniae*, two research groups, one in Belgium and one in Germany, attempted the genetic modification of plants with DNA in the late 1960s and early 1970s. At that time, no living cells of any kind, aside from only three bacterial genera, had been shown to incorporate exogenous (externally supplied) DNA. It should be remembered that DNA is a very large molecule that scientists thought would not be able to cross cell membranes. The three types of bacteria able to take up DNA spontaneously were regarded as strange exceptions, and nobody even dreamed that eukaryotic cells, such as plant cells, for example, would be capable of doing the same.

Lucien Ledoux of the Belgian Nuclear Study Center and Dieter Hess of Hohenheim University in Germany thought otherwise. Both viewed the idea of cells' DNA impermeability as false dogma. They thought that plant cells should be able to import DNA, and they set out to put their new hypothesis to the test. The future would show that they were right, but for all the wrong reasons. Ledoux was a rather corpulent gourmet already in his forties when his work on DNA uptake in plants fully developed. He was fond of the authority he held over people in his lab and did not encourage critical discussion. Hess looked much more ascetic and seemed to be imbued with quiet arrogance. He mostly worked alone, contrary to Ledoux, who loved a good entourage. (The photo on page 67 shows some of the scientists involved in the controversial aspects of DNA uptake by plants as published in the early 1970s.) These two researchers managed to open a large can of worms that kept many scientists busy for years. Their experimental results seemed to confirm each other's findings and were published practically in parallel. However, as we will see, they did not interpret their results correctly. Let us see what they did and found.

There are basically two ways to determine whether or not some molecules are taken up by cells. The first method is to make these molecules radioactive, incubate the cells with them, and see whether or not the radioactivity is incorporated. The other way is to see whether these molecules have any biological effect on the treated cells. In the case of DNA, this biological effect would be expected to be hereditary. Ledoux and his coworkers used both approaches, whereas Hess restricted himself largely to studying the biological effects of DNA. Let us start with results obtained by the Ledoux group, which published first in this area.

Most of the work done by this group consisted in soaking dry seeds in solutions of radioactive bacterial DNA and water. The rationale was that the seeds, which swell when put in contact with water, would absorb the DNA molecules dissolved in it. As the seeds germinated and the seedlings developed, samples were taken, ground up, and processed for DNA extraction. This DNA was then analyzed for radioactivity content. What results did they expect from such experiments? One would think that only two possible results could be obtained. Either the seeds would not take up any foreign DNA (possibly because DNA is too large to

Members of the Ledoux laboratory in 1974. From left to right: Pol Charles, Max Mergeay, Andy Kleinhofs (a visitor from the United States), Paul Lurquin, Raoul Huart, and Lucien Ledoux. At that time, Kleinhofs, Lurquin, and Mergeay had started questioning the scientific validity of the first plant genetic-engineering experiments performed in Mol. (From Lurquin, P. F. 2001. *The Green Phoenix: A History of Genetically Modified Plants*. New York: Columbia University Press. Figure 1.5. Reprinted by permission of the publisher.)

be incorporated) and would not accumulate any radioactivity, or DNA would be able to penetrate the germinating seeds and perhaps even find its way inside the dividing cells of the seedlings. In that case, radioactive foreign DNA would be found intracellularly or perhaps in the spaces that separated cells. Experimental results showed that radioactive DNA was indeed incorporated by germinating seeds of two test organisms, barley and *Arabidopsis thaliana*. This finding was encouraging.

The next problem, then, was to address the question of the physical nature of the radioactive foreign DNA found inside the plants. Here also, one could envision two possibilities: Either this DNA may reside inside the cells or intercellular spaces for prolonged periods of time, or, alternatively, this DNA would be rap-idly degraded by plant enzymes. (These DNA-degrading enzymes from plants should not be confused with the bacterial restriction enzymes encountered earlier.) Plant enzymes degrade DNA by cutting it randomly, not at specific sites as bacterial restriction enzymes do. The result of this degradation in plants would be the production of radioactive DNA building blocks that would be used by the plant cells to make more of their own DNA.

If the foreign DNA is left unscathed, all the recovered radioac-tivity should be associated with it. If, on the contrary, the foreign DNA is degraded and reused for plant DNA synthesis, all the radioactivity should be associated with the plant DNA. When DNA-treated barley and *Arabidopsis* seedlings were analyzed, neither result was obtained. The radioactive DNA molecules behaved as if they consisted of stretches of foreign bacterial DNA physically joined to the resident plant DNA, that is, integrated within it. These researchers claimed to have observed a recombi-nant DNA-type of phenomenon, occurring in plant cells, several years before the Cohen-Boyer in vitro experiment. These conclu-sions must have been exciting, surely; but were they warranted? It should be remembered that in those days, cloning had not yet been invented and that *total* bacterial DNA, not cloned genes, was used in these experiments. Further, the DNA analysis tech-nique used by the Ledoux group was rather crude, and the results did not convince everyone. Still, several investigators, attracted by the possibility of transferring foreign genes to plants in such a simple way, set out to verify these conclusions.

When their results were assembled, they showed that these plant DNA/bacterial DNA hybrid molecules were nowhere to be seen. After short incubation times, donor bacterial DNA could be found associated with the plant tissues, but, as time went by, the bacterial DNA was progressively degraded and reused for plant DNA synthesis. This must have been very discouraging because degraded foreign DNA cannot contain intact foreign genes.

Many investigators spent years trying to duplicate Ledoux's foreign DNA integration results in plants and failed, except under one set of specific conditions: when seeds were contaminated with bacteria.

All plant seeds are naturally associated with a variety of microorganisms that do not usually interfere with the germination and growth processes. It is possible to sterilize seeds with chemical agents, but this technique is not always perfect. Residual bacteria can easily hide in tiny cracks of the seed coat and escape sterilization. Keeping in mind that bacteria divide on the average at least ten times faster than plant cells, it becomes easy to understand what happened in the Ledoux experiments. As the radioactive foreign DNA was degraded by plant cells, and perhaps by contaminating bacteria themselves, radioactive building blocks were released. These building blocks were then rapidly used by the fast-growing, contaminating bacteria to make their own DNA. Thus, the foreign bacterial DNA never became integrated within the host plant DNA. Live bacterial contaminants of the seeds only made it look that way.

It took roughly seven years, from 1968 to 1975, to understand the experimental artifacts and realize that plant genetic engineering via DNA uptake by seeds and tissues was not exactly imminent.[1] In the meantime, Ledoux and Hess had also published intriguing results indicating that externally fed DNA could change the genetic properties of recipient plants. These results seemed to support the idea that, indeed, plant seeds and tissues could pick up and maintain externally supplied DNA. Now, if the results with radioactive bacterial DNA were really due to contamination, how could one explain the biological effects of DNA that was never kept intact in the first place? Either the proponents of DNA uptake and expression by plants were right and the detractors were wrong, or the reverse was true.

Dieter Hess, meanwhile, had published several articles in which he claimed that flower color in petunias could be influenced by DNA uptake. For this, he treated white-flowering (that is, flowers that make no pigments) petunia seedlings with DNA extracted from a red-flowering variety (flowers that synthesize a red pigment). Some of the treated seedlings developed into plants bearing red or reddish flowers and sometimes even developed into plants bearing flowers containing the red pigment concentrated in sectors. Further, crosses with these plants showed that flower color had been inherited through successive generations.

However, there was a big problem—the laws of Mendel were not obeyed in these crosses. I explained earlier how Mendel had derived his laws of genetics from the quantitative behavior of offspring of crosses. Recall that phenotypic categories (such as red-flowering versus white-flowering, for example) came in precise ratios, such as 3:1. Hess did not observe this, his ratios were non-Mendelian. In fact, his results defied interpretation and could not be attributed directly to a genetic effect of the donor DNA that contained genes responsible for the synthesis of the red pigment. We know now that flower pigment synthesis in petunias is extremely sensitive to environmental conditions, such as light intensity, soil acidity, temperature, and infection by viruses. In all likelihood, Hess's results were due to such conditions, not to the DNA he used to treat his plants.

Lucien Ledoux's group performed somewhat similar experiments, this time using *Arabidopsis* and bacterial DNAs as donors. Remember that this group had used *Arabidopsis* to claim integration of bacterial DNA, after uptake by seeds, within the host plant DNA. Since they obviously believed in their own claims (now known to be wrong), they set out to test whether foreign DNA taken up by seeds showed any biological activity. Instead of flower color as a marker for the effect of donor DNA, they used *Arabidopsis* mutants that were unable to make the essential nutrient thiamin (vitamin B1). These mutants are genetically blocked in the metabolic reactions that make thiamin and cannot survive without an outside source of this compound. Researchers grow and propagate these mutants simply by adding a thiamin solution to growth medium.

The logic of the Ledoux experiments was as follows: Since bacterial DNA, as they thought then, can be taken up by seeds and become integrated within the plant genome, it might be able to correct the genetic defects of the plant mutants, *provided* it contained the relevant thiamin genes. Then, integrated bacterial genes that would direct the synthesis of thiamin would supplement the defective plant genes, and the treated mutants would become thiamin-independent. It was known at the time that many types of bacteria produce thiamin to survive and divide, and they contain the genes needed to manufacture it. Thousands of mutant *Arabidopsis* seeds were treated with bacterial DNA known to harbor thiamin genes and, sure enough, a significant fraction of the treated seedlings grew into thiamin-independent individuals. The conclusion seemed obvious: Bacterial DNA containing thiamin genes had been taken up by the plants and had corrected the genetic defect. But here again, was this really the case?

The concept of DNA uptake and integration through seed soaking had been heavily criticized and shown by others to be no more than an experimental artifact. How then was one to interpret the biological effects of DNA on thiamin mutants reported by the Belgian group? In science, it is imperative that experimental results published by one researcher or research team be exactly reproducible by others, provided they use the same material and the same experimental conditions. Was this the case with the *Arabidopsis* correction experiments? The answer is no. Others could not confirm the effect of bacterial DNA on thiamin mutants. Worse, in this case as well as the one mentioned earlier, the *Arabidopsis* mutants seemingly corrected by bacterial genes did not follow Mendel's laws in crosses. Again, as in Hess's case, expected Mendelian phenotypic ratios were not observed. What was happening? Others solved the problem after requesting and receiving corrected seeds from the Ledoux lab. Their conclusions were devastating: The genetic effects seen by the Ledoux group were not due to DNA; they were due to seed stock contamination! In other words, the little bags used to store *Arabidopsis* seeds must have contained mixtures of mutant and normal seeds and even a special type of mutant seeds that reportedly had never been used in this laboratory. Again, there was no solid evidence that externally supplied DNA had any biological effect on plants.

We know now that these experiments could not have worked; the bacterial genes used were evidently equipped with their natural bacterial promoters, which do not function in plants. This, however, was not known in 1974, the year of publication of the Ledoux report, and that is why others had to spend a considerable amount of time verifying results that were impossible to achieve in the first place.[2]

By the mid-1970s the field of plant transformation was in shambles. All reports of foreign DNA uptake, integration, and genetic expression in plants had been contradicted. Bitter arguments took place between supporters of the idea and its detractors. There is no doubt today that the supporters were wrong, but, at the time, plant scientists who were not involved in DNA uptake experiments began to rail against the whole endeavor. They became convinced that these plant genetic engineers did not know what they were doing, and, what was more, their bickering was often undignified.

It was not only Ledoux and Hess who made wild claims about plant transformation. An Australian group announced at about the same time that bacteriophage genes could be expressed in plant cells. This claim was quickly refuted by a British team who could not reproduce these results. Remember that bacteriophage genes are designed to function in bacteria, not in plants, and their promoters are, in fact, inactive in eukaryotes. This too was unknown when the experiments with bacteriophages were done. Basically, the field of plant transformation had evolved into a tangled mess of claims and counterclaims. Dissenting scientists managed to stop short of insulting one another in public, but tempers ran high. This made for lively scientific meetings (where red faces were common), but it did not advance knowledge. Some scientists became discouraged by this state of affairs and simply abandoned the field. Others started thinking about different ways to introduce DNA into plant cells.

An Unlikely Ally

Plant genetic engineering was finally made possible via a totally indirect route, the study of the plant disease called *crown gall*. Crown gall disease is characterized by tumor formation at the crown of the plant, the area of the stem that is closest to the soil. This disease had been known since the early 1900s, and, because of its fairly minor impact on agricultural production, was studied by just a handful of scientists. Little did anybody know that the elucidation of this disease would lead directly to applied plant genetic engineering. Crown gall-diseased plants are easy to recognize. They grow tumors on their stems, as shown in Figure 4.1. In fact, crown gall disease is a genuine plant cancer, as the tumors continue to proliferate until the plant dies. These tumors can be excised from the affected plants, propagated in the lab on nutrient medium, and studied. Today, we know that crown gall tumors are completely unrelated to human cancer.

Researchers discovered early that crown gall tumors are incited by a bacterial pathogen, *Agrobacterium tumefaciens*, which is a normal inhabitant of soils. When a plant is infected by *A. tumefaciens*, usually through some type of wound, some of its cells turn into tumor cells. To put it differently, *Agrobacterium* somehow instructs plant cells to lose their identity and become undifferentiated tumor cells. It's as if this pathogen reprograms the normal plant cells to change into cancerous tumor cells. How can a bacterium possibly do that? In addition, crown gall tumor cells are very peculiar; they can be grown on artificial medium without

FIGURE 4.1   Crown Gall Tumor (white, elongated, somewhat fuzzy mass) Growing on the Stem of a Tobacco Plant Inoculated with A. *tumefaciens*. Several leaves have been clipped off to reveal the tumor.

phytohormones. Phytohormones are plant hormones that are normally produced by plants in nature. However, plant tissues cultivated in petri dishes containing artificial nutrient medium require an outside source of phytohormones and will divide only if experimenters add these hormones to the growth medium. How did crown gall cells lose their dependency on phytohormones? It turns out that they actually produce their own phytohormones. But then, how do crown gall cells do this when normal plant cells cultivated under the same circumstances do not?

Next, crown gall cells were found to synthesize unusual amino acids, such as nopaline and octopine (so named because it had previously been discovered in octopus!). These amino acids are not used for protein synthesis. What are they doing there and why is it that normal plant cells never manufacture them? Finally, once *Agrobacterium* has infected a plant, thereby causing crown gall tumors to appear, the bacterial cells are no longer necessary to maintain the tumorous state. In other words, *Agrobacterium* is necessary to trigger the disease, but once it has done so, it can be eliminated from the tumor, which continues to grow and maintains its undifferentiated state. How can *Agrobacterium* possibly do so much to change plant cells? All these observations and questions were in place by the early 1950s and were generated largely by the work of two research teams, Armin Braun's in the United States and Georges Morel's in France.

To summarize, crown gall cells are stuck in an undifferentiated tumorous state, produce their own phytohormones, manufacture unusual amino acids, and do not need the causative agent, *Agrobacterium*, to maintain their bizarre properties. Since all these features are very stable, it is reasonable to assume that a permanent change at the DNA level has taken place in these cells. Also, *Agrobacterium* is directly responsible for inducing these changes. How? Could it be that somehow *Agrobacterium* is able to mutate plant cells? This was hard to accept since the bacteria would have to always mutate the same genes. This consistent change does not happen with real mutagenic agents like X-rays and chemical mutagens that randomly mutate all possible genes. The hypothesis of *Agrobacterium* as an agent inducing mutations in plants quickly was abandoned.

Then, some researchers started toying with the idea that crown gall tumors were produced by *DNA transfer* from *Agrobacterium* to plant cells. This idea was put forth because sometimes plant and animal hybrids display new phenotypes, new biological properties. Could it be that somehow *Agrobacterium* donated its own genes to plant cells and that these genes were responsible for the crown gall phenotype? This idea was debated more than once and almost invariably rejected because it seemed to be so crazy. Indeed, this would have meant that a bacterial cell was in some way able to mate with a plant cell. This sounded ridiculous and was even heretical because it was tantamount to claiming that, for example, *E. coli* and dog cells could mate, which of course they cannot. Only viruses were known to introduce their genes into plant cells, not bacteria. Then, what on earth was this activity, this tumor inducing principle (TIP), as it was later called?

Spurred by the weird idea that maybe *Agrobacterium* could in fact introduce its genes into plants, some researchers tried to induce crown gall by incubating plant cells with purified *Agrobacterium* DNA, which obviously contains all *Agrobacterium* genes. They claimed it worked: According to these investigators, purified *Agrobacterium* DNA was able to transform normal plant cells into crown gall cells in the laboratory. This success was of short duration. Once again, as in the case of Ledoux's experiments seen earlier, these results could not be reproduced independently. Then, *Agrobacterium* RNA was tried (remember that RNA is a copy of DNA), and it worked too! Alas, this was yet another artifact that could never be reproduced by others. And still more weird science was to come.

It was known in the 1970s that *Agrobacterium*, like *E. coli*, plays host to some bacteriophages. Sometimes bacteriophages are not lethally virulent and can reside inside bacterial cells without killing them. *Agrobacterium* was a known host to this type of bacteriophage. Could such a mild bacterial virus be the TIP that theoretically could be transferred from *Agrobacterium* to plant cells through some kind of mating? Two research groups claimed that crown gall cell DNA contained bacteriophage DNA. Again, this claim was dismissed as wrong, irreproducible. All this squabbling happened as others reported on equally wrong uptake, integration, and biological effects of foreign DNA in plants. All the false observations described herein mutually reinforced one another, and this made the life of skeptics miserable because, as we now know, two wrongs did not make a right. Crown gall research, like early attempts at genetically engineering plants, was getting a bad reputation as the domain of scientists whose results could not be reproduced. Sloppy laboratory procedures were responsible for this sad state of affairs. This would not last, fortunately, because it was soon discovered that *Agrobacterium* indeed *is* able to transfer its genes to plant cells in a way that nobody could have predicted.

Agrobacterium Does Transfer Genes to Plants After All

Reason returned to the field of crown gall research through the study of the causative agent, *Agrobacterium,* and not the study of the tumors themselves. As we examined earlier, many types of bacteria harbor minichromosomes, or plasmids. These plasmids were a hot topic of research in the early 1970s, not only because they were used in the first gene cloning experiments, but because they were also biologically interesting. Among other things, they imparted antibiotic resistance to bacteria (a serious problem in human medicine). A group at the University of Ghent, Belgium, started exploring the possibility of finding plasmids in *Agrobacterium tumefaciens* and related species. Jef Schell and Marc Van Montagu of the Ghent team indeed did discover very large plasmids, 200,000 base pairs (bp) long, in *A. tumefaciens.*

What was really exciting was that avirulent *Agrobacterium radiobacter*, a close relative of *A. tumefaciens* that does *not* cause crown gall, did *not* contain such a plasmid. Further research showed that the virulence of *A. tumefaciens* could be abolished by growing the bacteria under conditions (at sublethal temperature, for example) that caused the loss of plasmid DNA. Therefore, plasmidless *A. tumefaciens* could no longer incite crown gall in plants. Finally, *A. tumefaciens* strains made avirulent by loss of plasmid could be restored to a fully virulent state by reintroduction of the large plasmid.

At last, logic prevailed; it was now clear that whatever caused tumor formation in plants was directly linked to the presence of a large plasmid in *Agrobacterium*. This conclusion was rapidly confirmed (a happy change) by a research group at the University of Washington, Seattle. The plasmid is now called pTi or Ti plasmid, where p stands for plasmid and Ti for tumor-inducing. Thus, pTi held the key to an understanding of crown gall. But how did it act? Was pTi transferred to plant cells after some kind of mating with *Agrobacterium*? Another U.S. group quickly published results showing that, indeed, the complete Ti plasmid was present in the DNA of crown gall cells. This report was just as quickly shown to be wrong.

At this point, in 1976, a fantastic breakthrough occurred. The University of Washington team was led by Mary-Dell Chilton (pictured in the photo on page 79), Eugene Nester, and Milton Gordon. All three were well aware of the numerous and unconvincing attempts made by others to introduce foreign genes into plants. In fact, Chilton herself had helped debunk the claims made by Ledoux. Nester and Gordon had even visited the Ledoux laboratory and a Dutch lab from which claims of bacteriophage DNA presence in crown gall tumors had originated. Clearly, they were all very interested in the prospect of transferring foreign DNA to plants, but they also knew that so far, this phenomenon had not been demonstrated convincingly. Once they disproved

Dr. Mary-Dell Chilton, the lead
investigator of the University of
Washington, Seattle group, that
solved the crown gall riddle.
(Courtesy of Dr. M.-D. Chilton,
Novartis, North Carolina).

the claim that the whole pTi was present in crown gall, they wondered if perhaps, a *portion* of the Ti plasmid was present in these
cells. This would still imply that a bacterial cell could transfer
some DNA to a plant cell—a heretical idea—but the time was
now ripe to test a hypothesis that seemed profoundly far-fetched.

The Seattle team used a sensitive technique, called *DNA
hybridization*, to check crown gall DNA for the presence of pTi
sequences. For this, they cut pTi with a restriction enzyme, isolated the fragments, and probed crown gall DNA with them. There
it was: One of the pTi fragments gave a strong positive signal. The
unthinkable had been demonstrated: *Agrobacterium was* able to
transfer some of its genes, by way of a fragment of the Ti plasmid,
to plant cells. This fragment is small, about 20,000 base pairs,
roughly one-tenth of the plasmid length, and represents but
0.0011 percent of the total DNA present in crown gall cells. Only
the use of a hypersensitive technique and rigorous lab practices
had allowed this discovery. Finally, the TIP had been identified,

and, at last, the discovery was quickly confirmed by several independent research groups.

Now it was demonstrated: Crown gall is caused by a bacterium, *A. tumefaciens*, that is able to insert a piece of its DNA (called *T-DNA* for transferred DNA), big enough to harbor about twenty genes, into the genome of a plant cell. The next step was to figure out what these bacterial genes did once they were transferred. Did they integrate inside plant genes, thereby disrupting their functions and causing the crown gall phenotype? This did not appear to be terribly plausible since crown gall cells produce their own hormones in addition to rare amino acids. These two phenotypes are unrelated, and they correspond to *gain* of function (gain of genes if you will) by plant cells, not *loss* of function—an activity that would accompany disruption of resident genes by random integration of *Agrobacterium* DNA. It was hypothesized subsequently that T-DNA genes were probably transcribed and translated in plant cells, in spite of their bacterial nature.

This was 1977, a year that saw the beginning of a fierce competition among three laboratories: the Ghent group, the Seattle group, and a group located at the University of Leiden in the Netherlands. This latter group was headed by Rob Schilperoort. These three research teams would monopolize the field for several years.

After six years, the crown gall riddle was solved completely, once and for all. First, the T-DNA genes transferred by *Agrobacterium* were shown to be transcribed in crown gall cells, meaning that an RNA copy of these genes was produced in plant cells. Next, it was demonstrated that T-DNA from *Agrobacterium* was physically integrated within the plant genome. Finally, it was shown that the T-DNA genes coded for the synthesis of the amino acids octopine or nopaline *and* for the production of phytohormones. These features explained the mysterious behavior of crown gall cells, phytohormone-independence and synthesis of unusual amino acids. These properties were due to the acquisition of bacterial genes by plant cells. This also meant that what previous researchers had tried (and failed) to achieve had been done all along by a lowly bacterium, *Agrobacterium tumefaciens. Agrobacterium transfers its DNA to plant cells and is a natural genetic engineer.*

It may be wondered at this point how it is possible that genes of bacterial origin can be expressed in plant cells. I have explained earlier that this is not to be expected because of the incompatibility of gene promoter regions. It turns out that T-DNA genes are not typical bacterial genes; their promoter and terminator regions are eukaryotic and well designed to function in plant cells. In fact, and for all practical purposes, T-DNA genes are not expressed in *Agrobacterium* itself. How then did these eukaryotic-like genes end up in a bacterium? Nobody really knows, but it is not impossible that the phytohormone genes were captured by *Agrobacterium* from plant cells as a result of some evolutionary pressure. As for the genes coding for the synthesis of nopaline or octopine, their origin is even more obscure since no known contemporary plants normally produce these amino acids. And why both crown gall tumors and octopus synthesize octopine is anyone's guess.

It should be noted that the type of genetic engineering that *Agrobacterium* is capable of performing is, of course, of no practical interest to humans. After all, plant tumors present no danger nor do they possess any special nutritive value for humans. Further, the human body has no use for octopine or nopaline. The point, however, is not to marvel at what *Agrobacterium* can do (although that is interesting in its own right), but rather to determine whether or not T-DNA can be used as a vector to introduce genes of interest into plants. That is, if *Agrobacterium* can transfer its T-DNA to plant cells, could it be that other genes, incorporated within the T-DNA by recombinant DNA technology, would also be transferred to plant cells? The answer turns out to be yes.

This process was first demonstrated by cloning an antibiotic resistance gene inside the T-DNA and allowing *Agrobacterium* so engineered to infect plants. The infection produced crown gall tumors that were resistant to the antibiotic (shown in Figure 4.2). Why did the researchers use as a foreign gene one that determines resistance to an antibiotic? The point here was to demonstrate the *feasibility* of foreign gene transfer via *Agrobacterium* T-DNA. Plant cells are sensitive to certain antibiotics that, when supplied at high enough concentrations, will kill them. If an antibiotic-resistance gene is introduced into plant cells, and if this gene is expressed, the transformed plant cells will survive in the presence of a lethal dose of the antibiotic. Thus, the researchers exposed the engineered crown gall tumors to high doses of antibiotic, and the tumors survived. This showed that the antibiotic-resistance gene incorporated into the *Agrobacterium* T-DNA had not only been transferred to plant cells, it was expressed correctly in the new plant host. These experiments also demonstrated that cloning a totally foreign gene (the antibiotic-resistance gene) inside the T-DNA did not interfere with its transfer.

This work was done in 1983 in parallel by three different teams. That year marks the true birth of plant genetic engineering. These three teams were the University of Ghent–Schell-Van Montagu group, the Washington University, St. Louis, group headed now by Mary-Dell Chilton, and a group based at Monsanto, the industrial giant. The appearance of the Monsanto team among basic research endeavors demonstrated great foresight on the part of its industrial managers. (In 1983 plant transformation had not yet taken on an applied aspect.) At that point, no patents had been granted to anybody, simply because it was not yet an accepted norm (as it is today) for academic scientists to file patent applications for their discoveries. Monsanto's participation in fundamental crown gall research would allow the company to claim precedence in the field of plant genetic engineering.

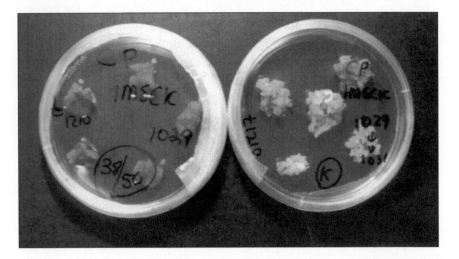

FIGURE 4.2   Experiment Demonstrating *Agrobacterium*-mediated Antibiotic-resistance Gene Transfer into Plant Cells. Left: control leaf pieces put in a Petri dish containing medium supplemented with the toxic antibiotic kanamycin. The leaf pieces turn brown after a few days and show no cell proliferation. Cells in these explants are dead. Right: leaf pieces that were treated with *Agrobacterium* whose T-DNA was engineered with a kanamycin-resistance gene. The medium in this Petri dish also contains a lethal dose of the antibiotic. This time, however, abundant cell proliferation is observed on the explants. These cells have become resistant to the antibiotic and multiply in its presence. The inscriptions on the Petri dish lids identify the experiment in laboratory shorthand.

The Taming of *Agrobacterium*

As I explained, foreign gene transfer mediated by *Agrobacterium* T-DNA was solidly established by 1983. As great as this achievement was, it resulted only in the production of genetically engineered crown gall tumors, not fertile green plants. Even assuming that nutritionally important genes could be transferred to plant cells in that way, human consumers would not be thrilled to eat their steak or salmon accompanied by boiled (or even stir-fried) crown gall tissues. One needed a way to make T-DNA nononcogenic, that is, one needed to understand which genes harbored by T-DNA caused tumor formation in order to eliminate them. The T-DNA was fully sequenced and three genes coding for the production of phytohormones were identified. These three genes are responsible for the production of an auxin and a cytokinin, two natural plant hormones. What is remarkable is that these phytohormones were produced in precisely the amounts necessary to keep crown gall cells in an undifferentiated, tumorous state. The reasoning then was that if one could eliminate these three genes from the T-DNA, tumor formation would be avoided. Of course, one would have to ascertain that T-DNA would still be transferred to plant cells if the phytohormone genes were eliminated.

This turned out to be the case; the hormone genes, also called *oncogenes* or *onc* genes, have nothing to do with the mechanism of T-DNA transfer. The genes whose function it is to clip the T-DNA out of the Ti plasmid are located elsewhere on pTi itself. They are called *vir* genes, short hand for *virulence genes*, because when they are inactivated, the T-DNA cannot be transferred from *Agrobacterium* to plant cells. By themselves, *vir* genes do not cause crown gall; they simply code for enzymes that process the T-DNA. Thus, the taming, or disarming, of the Ti plasmid, the process that destroys its ability to cause tumors in plants, should be the result of simple oncogene removal from the T-DNA. Remarkably, disarming the T-DNA was achieved in 1983, by the same three teams listed earlier. The predictions were correct; removing the *onc* genes from the T-DNA and replacing them with an antibiotic-resistant gene resulted this time in the formation of green, fertile plants that were, in fact, resistant to the antibiotic after *Agrobacterium*-mediated gene transfer. This was a breakthrough: *Agrobacterium* had been engineered in such a way that it had kept its ability to transfer DNA to plant cells while losing its ability to cause tumor formation (as shown in Figure 4.3).

Plant genetic engineering was, from this moment forward, a full reality. The chase was over; these results were for real. All these basic discoveries led to a widely used technique to transfer

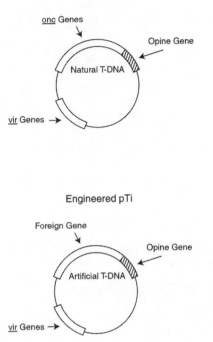

Normal pTi

onc Genes

Opine Gene

Natural T-DNA

vir Genes →

Engineered pTi

Foreign Gene

Opine Gene

Artificial T-DNA

vir Genes →

FIGURE 4.3   Structures of a Normal, Virulent pTi Molecule and Its Nonvirulent, Engineered Counterpart

foreign genes to plants. This technique is now known as *cocultivation with Agrobacterium* or *Agrobacterium-mediated gene transfer*. In it, pieces of plant material (often leaf discs obtained with a single-hole paper punch) are simply incubated for a few hours with *Agrobacterium* containing an engineered T-DNA. Subsequently, these leaf discs are placed on medium that contains chemicals to trigger plant regeneration from leaf cells, and the regenerated plants are analyzed for foreign gene presence and expression.

With the verification and commonplace use of this procedure in place, the story of the fundamental discoveries about the nature of crown gall disease came to an end. See Figure 4.4 for a simplified summary of what we know about the mechanism of DNA transfer from *Agrobacterium* to plant cells. Basically, the T-DNA leaves the Ti plasmid, travels from *Agrobacterium* to the

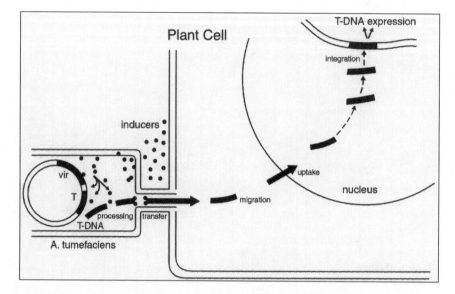

FIGURE 4.4   Simplified Representation of Gene Transfer Between A. *tumefaciens* and a Plant Cell. The plant cell (schematized on the right) excretes small molecules (dots) that penetrate the A. *tumefaciens* cell (on the left) and activate the *vir* genes.

plant cell through a tunnel made by the bacterium itself, where it penetrates the plant cell, then its nucleus, and gets integrated within the plant DNA. Once there, it expresses its phytohormone and opine producing genes. The mechanism is the same when the T-DNA is disabled by oncogene removal and engineered with a foreign gene. As soon as it was demonstrated that *Agrobacterium* and its disarmed pTi were excellent vehicles to transfer test genes (antibiotic-resistance genes), there was an enormous proliferation of scientific articles that extended the utility of this technology. Many articles dealt with basic plant biology, but others were clearly of an applied nature. Applied plant biotechnology had begun.

For their crucial discoveries, Mary-Dell Chilton, Gene Nester, Marc Van Montagu, and Jef Schell were elected to the U.S. National Academy of Sciences.[3]

Back to the Beginning

Admittedly, the process of *Agrobacterium*-mediated gene transfer to plants is not simple. Technically speaking, though not terribly complicated, this technology requires the knowledge of some microbiology and some experience with cloning and simple plant tissue culture techniques. This was not to everyone's taste, and scientists still hoped one day to achieve direct DNA transfer. Strange but true, *Agrobacterium*-independent DNA transfer had not been abandoned completely. Everyone knew that a Ledoux or Hess approach to plant transformation with naked DNA was not worth pursuing. Yet the question persisted: Were there other possibilities?

In the 1960s scientists learned to manipulate single plant cells in laboratory petri dishes. They incubated pieces of leaves in a solution of enzymes able to digest the thick cellulosic cell wall that surrounds all plant cells. The result of such a treatment was a suspension containing millions of what are called *protoplasts*, single plant cells without cell walls. Provided with appropriate nutrients, these protoplasts could regenerate a cell wall and start dividing. Lumps of newly formed cells could even be regenerated into whole, fertile plants (see Figure 4.5). Protoplasts looked like a clean and attractive system for DNA uptake studies when compared to anatomically complicated seedlings or mature plants. Sterility of protoplast suspensions, that is, absence of contaminating bacteria, was also much easier to control. (Remember that bacterial contamination had confused early DNA uptake experiments.) Further, protoplasts were devoid of a cell wall, a possible barrier to DNA uptake. By the mid-1970s, DNA uptake studies with protoplasts were in full swing. Labs in Belgium, Canada, Japan, The Netherlands, the United Kingdom, and the United States seemed to agree that protoplasts would be *the* tools that would show plant genetic engineering to be possible. As we know, however, that did not happen; *Agrobacterium*-mediated gene transfer came first. Nevertheless, protoplasts finished a very close second.

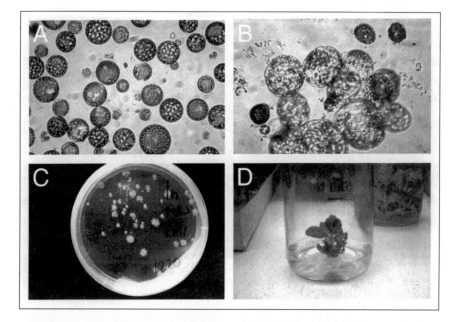

FIGURE 4.5   Regeneration of Plants from Protoplasts. (a) Photomicrograph of protoplasts freshly isolated from leaves. The protoplasts are spherical due to osmotic pressure and can be seen to contain chloroplasts. (b) After a few days, protoplasts make a new cell wall, lose their perfectly spherical shape and start dividing to form small clumps of cells. (c) After about a week, the cell clumps have grown and can be plated in a petri dish where they continue to grow. The microscopic clumps then yield masses of cells that can readily be seen with the naked eye and be picked individually. (d) Individual clumps can then be transferred to special growth medium where they regenerate green plants.

It was demonstrated that protoplasts could take up DNA from their growth medium if treated with certain chemical helpers designed to promote permeability of the protoplast membrane. Not only that, purified pTi DNA isolated from *Agrobacterium* could be taken up by protoplasts that, when allowed to divide, gave rise to crown gall tumors growing in vitro, just as live *Agrobacterium* cells did in plants. The tumors so obtained contained integrated T-DNA, produced their own phytohormones and opines (octopine or nopaline), and were indistinguishable from crown gall tumors produced in nature. Thus, *Agrobacterium* was not necessary to achieve transformation of plant cells. Its Ti plasmid DNA alone could do the same thing when chemical helpers allowed it to cross the protoplast membrane. By 1982 there was no doubt that pure DNA could transform plant protoplasts. However, by then, only crown gall transformation had been achieved. In 1984, only one year after the triumphal announcement of *Agrobacterium*-mediated plant transformation, the same goal was achieved with protoplasts and naked, non-pTi DNA. The transforming DNA was simply an antibiotic resistance gene located next to a plant-expressible promoter and cloned in an *E. coli* plasmid vector. This DNA also integrated itself within the recipient protoplast genomes, meaning that the concerns expressed in the early 1970s about nonreplication of bacterial plasmid DNA in plant cells were unfounded. Foreign genes cloned in bacterial plasmids did not need to replicate in the plant environment, they became physically integrated within the plant DNA, just like *Agrobacterium* T-DNA. Therefore, what *Agrobacterium* could do, humans also could do. This remarkable result was achieved in a laboratory in Basel, Switzerland, under the direction of Ingo Potrykus (pictured on page 90), a congenial fellow who sported a fierce black (at least it was in 1984) beard that made him look like a debonair Caribbean pirate or perhaps a more serene version of Melville's Captain Ahab.

It is interesting to note that in many ways, the transformation of plant protoplasts with recombinant DNA was conceptually similar to Cohen and Boyer's experiment with the bacterium *E. coli*. First, the gene to be expressed in the recipient cells was cloned in a plasmid vector. Next, this vector was taken up by the host cells under conditions forcing the transfer of DNA across the cell membrane. The main difference was that the recombinant plasmid remained intact in *E. coli* and replicated independently, whereas, in plant protoplasts, the recombinant plasmid spontaneously integrated within the plant DNA and replicated along with it.

Dr. Ingo Potrykus, head of the research team that demonstrated direct gene transfer into plants. Dr. Potrykus is also one of the creators of "golden rice." (Courtesy of Dr. I. Potrykus, ETH, Zurich, Switzerland.)

Thus, there were in 1984 two parallel techniques for the genetic engineering of plants. The *Agrobacterium* approach required engineering of the Ti plasmid and a minimum amount of plant tissue culture. The direct gene transfer approach (using naked DNA and protoplasts) required a minimum of cloning expertise and a great deal of plant tissue culture experience. Since plant tissue culture is slow and expensive, the protoplast technique has practically been abandoned in favor of *Agrobacterium*. Yet the direct gene transfer methodology was not dead, because many researchers were reluctant to learn how to manipulate *Agrobacterium*. Oddly enough, direct gene transfer into plants was revived by the peaceful use of firearms. To help you understand this bizarre application, let me summarize the steps needed for plant genetic engineering.

What, indeed, is at the end of the line of plant genetic engineering? First, a cloned gene that will be expressed in a plant is needed. Next, a DNA uptake mechanism that introduces this gene into plant cells is needed as well. Finally, the plant cells so transformed must be regenerated into fertile plants that can be cultivated in the field and propagated. Typically, *Agrobacterium* or plant protoplasts can be used for gene transfer. What if, for whatever reason, one wanted to use neither *Agrobacterium* nor protoplasts? Necessity required that scientists invent a new technique that bypassed both *Agrobacterium* and protoplasts. Why not, then, try to force DNA into plant cells through *physical* means rather than through *chemical* means as in the case of protoplasts or with bacteria as in the case of *Agrobacterium*-mediated gene transfer?

This end is what a "gene gun" achieves; DNA is transferred through physical means using a gun. Its principle is the following: Microscopic gold or tungsten particles are suspended in a DNA solution, whereby some of the DNA molecules spontaneously stick to the metal surface. The metal particles are then harvested and placed in a device that accelerates them to high velocity, fast enough for the particles to penetrate plant cells, but not so fast that they stop once inside. Once there, the particles release their DNA, which ends up integrated in the host DNA. The original device was an actual handgun, firing genuine cartridges, but modified slightly for its scientific purpose. The main problems were, of course, safety and the inability to fine tune particle velocity. I saw one of these contraptions. It looked impressive; the barrel of the gun had been partly sawed off and welded to a lid that fit on top of a thick steel cylinder necessary to stop ricocheting metal particles. Smoke discharge was also a problem; it filled the lab and polluted the plant samples. Today, gun cartridges have been replaced by high pressure helium discharge devices. The technique works with plant embryos, a great advantage because plant embryos develop directly into whole plants. Alternatively, pieces of leaf material can be used, since fertile plants can be regenerated from leaf cells in many plant species.

Several types of transgenic plants have been produced with the gene gun, a technique now referred to as *biolistics*, a less colorful but more scientific-sounding term. Biolistics is relatively popular these days, especially with researchers who find it difficult to manipulate *Agrobacterium* with recombinant plasmids.

Finally, after the "shoot 'em up" type of approach, we reach the last direct gene transfer technology in use today, the "zap 'em" technique. Here, plant cells are suspended in liquid medium in the presence of DNA and subjected to brief electric shocks of a few hundred volts using equipment composed of a voltage generator, a capacitor to store electrical energy, an oscilloscope to monitor the discharge, and a small flask containing flat metal electrodes. The plant cell suspension is introduced into the flask, and the capacitor is charged and then discharged via the electrodes through the cell suspension. A typical discharge time is measured in several tens of milliseconds. The result is the microscopic opening of temporary holes in the cell surface through which DNA can travel. When the holes are resealed, which happens spontaneously, DNA is trapped inside and ends up integrated into the host genome, just as is accomplished with biolistics. The *electroporation technique*, as it is known, is still in development and may or may not become generally applied.

Interestingly, Chinese scientists are applying today a technique first developed in the United States some ten years ago, though it was never implemented. This technique is called the *pollen tube pathway*. When a pollen grain lands on the tip of the flower's female organ, called the *pistil*, it germinates and sends down a tube that eventually reaches the ovules, and the union of the male and female gametes occurs. The idea was to take advantage of the burrowing ability of the pollen to introduce foreign genes, via the tube, simply by putting a small drop of recombinant DNA on the tip of the pistil. This DNA then flows down the pollen tube and can be incorporated by the forming embryo. This technique originally was considered unreliable and was discarded. Chinese scientists have revived it, however, and their transgenic cotton plants, in particular, are reportedly produced that way.

To summarize, today there are two techniques very widely applied to achieve the transfer of DNA to plants: One is the *Agrobacterium*-mediated gene transfer and the other is biolistics. Of the two, academic scientists much prefer the first one, owing to its predictability and versatility—it works well with numerous types of plants. Corporate scientists are more divided; for them, patents and potential patent infringements are as great a concern as the feasibility of the techniques themselves. Regardless of the technique used, plant genetic engineering is now routine with many plant species.

To conclude, this chapter has used a historical timeline in its description of the progress of ideas in the field of plant transformation. In my opinion, this is the best way to explain how we got where we are and where we are coming from. The reader now knows that first attempts at genetically engineering plants are far from recent. Even though these first attempts in the late 1960s ended up in scientific disaster, the thought process leading to some of today's achievements is loosely based on the attempts made over thirty years ago. As with other scientific areas, the science of plant transgenesis has been cumulative and was punctuated by major breakthroughs that pierced the mist of uncertainty. Today, plant transformation is well established and practically routine. It should now be clear to the reader that plant transgenesis was developed to solve fundamental biological problems, not to jump immediately into profitable applications. The initial basic challenge was the transformation of plants with foreign genes. Applications were secondary.

Those who would like the public to believe that plant transformation is a mere flash-in-the-pan, concocted by supposedly conspiratorial scientists, are wrong. The problem lies in part in the poor communication skills of most scientists. As far as I know, practically all the people who have had any role to play in the development of plant transformation are still alive, and active, and mostly silent. Not once have I heard any of those whom I know personally claim that they were doing this research for pure profit. The transformation of plants with foreign genes was their Mt. Everest, and they climbed it because it was there. There was never any conspiracy by academic scientists to force genetically engineered corn or other vegetables down the throats of an innocent public. On the contrary, most scientists are using transgenic techniques in plants to study the basic mechanisms of plant life, not to produce supposedly monstrous organisms. It remains that the scientists who made possible the genetic manipulation of plants opened up an immense labyrinth whose complications we are still trying to fathom. What corporate scientists did with these basic discoveries, however, is another matter that will be examined in chapters that follow.

When I recall the competition between Ken Giles's laboratory and my own, described at the beginning of this chapter, I must admit that he won. We had both discovered in 1979 a promising technique to introduce DNA into plant cells. However, in 1981, his Ph.D. student, Steve Dellaporta, demonstrated that genes introduced by this technique were expressed in the recipient plant cells. My lab demonstrated this expression also, but only two years later. Ironically, nobody really won after all; our technique is now totally obsolete and has been replaced by much more efficient and simpler ones, such as biolistics and *Agrobacterium*-mediated gene transfer.

5

Plant Biotechnology:
Accomplishments and
Goals with Food Plants

D NA TRANSFORMATION TECHNIQUES are an invaluable
part of genetic technology. Without them, it would be
impossible to study the function of genes with the precision we
have today. Cloned genes can be manipulated at the DNA
sequence level and reintroduced into a natural or foreign host
and their behavior and activity monitored. Entire genetic pathways, collections of genes acting in concert, have been dissected
this way. However, the term biotechnology, of which DNA transformation is very much a part, conjures up the idea of applied science. After all, biotechnology is also a *technology*, a set of techniques. Biotechnology now impacts our daily lives, and it can no
longer be restricted to academic discussions. The applied nature
of biotechnology needs to be understood. However, applied science in the area of plant biotechnology did not come spontaneously and easily, as the following example shows.

I first encountered Ingo Potrykus, whom we met in the pre-
ceding chapter, in the mid-1970s, at the Max-Planck-Institute in
Ladenburg, Germany, near the city of Stuttgart. At that time he
was fully involved in experiments aimed at regenerating cereal

plants from protoplasts, a somewhat academic problem, in spite of the commercial aspects of cereals. This regeneration problem turned out to be an extremely difficult question, and Ingo was an unrelenting critic of half-baked data that claimed to have solved all the difficulties.

Without a doubt, cereals play a major role in human nutrition. Therefore, many laboratories were trying to understand the biology of these plants for both theoretical and applied purposes. For years, progress in this area was close to nonexistent. Remember that Ingo Potrykus was the leader of the team that first demonstrated direct gene transfer to plants in 1984. Since cereals were too difficult to manipulate in vitro, the team decided to use instead tobacco protoplasts, a well-established laboratory model system to investigate DNA uptake and expression in plants. But then, tobacco protoplasts were just a model system, not something that could immediately be applied to practical questions pertaining to human nutrition. Nevertheless, was Ingo Potrykus already thinking, in those days, about some kind of genetic modification of rice (a cereal) that could make life better for millions of undernourished children in the Third World? Perhaps. I really don't know for sure. At any rate, this is what he may have achieved today, provided his creation, "golden rice" (rice rich in provitamin A), becomes massively distributed to those in need. I will describe this rice variety later in this chapter. For the moment, suffice it to say that it took a full decade to learn how to engineer rice plants with *A. tumefaciens*. However, before the creation of golden rice, several other applications of plant biotechnology with crop plants have seen the light of day, and several types of engineered food plants are currently in full production.

Two main themes related to genetically engineered food plants can be distinguished:. crop protection and human nutrition, the latter being especially crucial in the developing world. Of all the applied aspects of plant biotechnology, crop protection through herbicide and insect resistance is by far the most developed and widely used in North America. In 2000, it was estimated that the United States accounted for 68 percent of the world-area planted with transgenic crop plants. Canada and Argentina accounted for 7 percent and 23 percent, respectively, whereas the rest of the world represented only 2 percent. China is the only developing nation where transgenic crops are cultivated on a significant scale, comprising 1 percent of total world-area. It was, in fact, China that inaugurated in 1992 the first planting of a commercial genetically modified (GM) plant, a virus-resistant tobacco. To date, 36 percent of the soybeans planted in the world are genetically modified, and so are 7 percent of the corn, 16 percent of the cotton, and 11 percent of the canola. It is estimated that about 60 percent of processed foods available in U.S. supermarkets contain a GM ingredient. Thus, GM foods can no longer be ignored; they are now omnipresent. The following sections explain how and why these GM plants were produced.

Crop Protection

Crop plants that grow in our fields are subjected constantly to real injury. The major culprits are plant pathogens: viruses, bacteria, fungi, and insects. The diseases they cause are known by common names often ending with *blight* or *rot*. One plant disease is even known as *take all*. Needless to say, farmers would be happier if these diseases could be controlled; and so would the consumer. Fewer losses to the farmer mean lower prices at the grocery store. An additional way to visualize crop protection is to think about herbicides. These chemicals are, of course, used to curb the growth of competing weeds. These weeds deplete yields just as surely as pests do.

Herbicides have been used with success by growers for decades. I say "success" and not "great success," not because herbicides don't work (they *do*), rather, I am taking into account the public perception of agrochemicals. Even though tempers flare occasionally when pesticide residues are detected in our fruits and vegetables, we must admit that pesticide use has not resulted in increased human morbidity. Nevertheless, public perception is important. For example, the discovery a few years ago of traces of the pesticide Alar® in apples produced in the state of Washington had a very negative impact on sales of apples nationwide. This alarm went off in spite of the fact that Alar® residues were present in extremely low amounts, well below human toxicity levels.

Many consumers have a favorable opinion of organically grown vegetables and fruits. But then, organic produce has not overtaken the market, perhaps because the public does not know what *organic* really means. And indeed, there is no agreement on what the term truly encompasses. Clearly, people are ambivalent about the use of chemicals on food they ingest, and rightly so, but the enormous majority of them have accepted the practice. At any rate, I think herbicides are here to stay. This trend exists simply because the ones in use today are safe and they keep prices low.

There is one big problem with herbicides, however: The really potent ones indiscriminately kill all plant life. These were (in pre-biotechnology days) called *preemergence herbicides* because they could be sprayed only before the seeds of crop plants germinated and emerged to grow above ground. However, the weeds that had already emerged would be killed. After germination, postemergence herbicides were sprayed to kill late-germinating weeds, or weeds whose seeds had been deposited later by the wind. Postemergence herbicides were not nearly as powerful as preemergence herbicides, simply because they could not be so toxic as to kill the valuable crop seedlings that had already emerged. Therefore, a combination of preemergence and postemergence herbicides led invariably to the use of herbicide cocktails, meaning greater cost and heavier chemical load in sprayed soils.

The application of biotechnology has changed that situation radically in two famous examples that concern the major broad-spectrum (kill-all) herbicides on the U.S. market—Roundup®, manufactured by Monsanto, and Liberty® (what a name for an herbicide!), manufactured by AgrEvo (now part of Aventis Pharmaceuticals). Both have excellent safety records. Biotechnologists working for these two companies reasoned that engineering crop plants for resistance to Roundup® or Liberty® would allow use of these herbicides as *both* preemergence and postemergence herbicides. This way, only *one* herbicide would be required, instead of two. Indeed, engineered crop plants would be insensitive to the herbicides at all development stages, from young seedling to adult. Weeds, on the other hand, would be killed, since they would always be sensitive to the broad-spectrum herbicide. This result was achieved with soybean and canola, among other widely farmed crops. In order to understand how this works, it is first necessary to understand how these herbicides kill plants.

Let us start with Roundup®. The chemical name of this herbicide is glyphosate. Glyphosate kills all plants by interfering with the function of an enzyme called *EPSP synthase*. The role of this enzyme is to make three essential amino acids—tryptophan, tyrosine, and phenylalanine—all needed for protein synthesis in plants. When glyphosate is present in a plant after spraying, EPSP synthase is inhibited, the three amino acids are not made, and protein synthesis is blocked. The plant dies as a result. It is important to note that humans *do not* possess the EPSP synthase enzyme, and, as a result, our protein synthesis mechanism cannot be inhibited by glyphosate.

Engineering plants for resistance to glyphosate was achieved when Monsanto scientists first screened in vitro-grown petunia cells for tolerance to glyphosate. Most cells died in the presence of the herbicide, but those that survived had a variant of the EPSP synthase gene whose gene product (the enzyme it codes for) was *less* sensitive to the herbicide. That gene was cloned. Next, that gene was recloned close to a very active promoter sequence. Recall that a promoter is necessary for the transcription of a gene. Not all promoters have the same affinity for the enzyme that performs transcription, RNA polymerase. Active promoters have high affinity for RNA polymerase and are thus transcribed abundantly. This way, many RNA copies of a gene equipped with an active promoter are made, then translated, and as a result, many copies of the protein are also made. The active promoter selected to direct transcription of the glyphosate-tolerant EPSP synthase gene was isolated from a plant virus, Cauliflower Mosaic Virus. This does not mean that engineered soybean plants contain a virus; only that a promoter sequence was cloned from it, not any coding sequence for the virus itself. Finally, this chimeric glyphosate-tolerant EPSP synthase gene was cloned in a disarmed T-DNA from *Agrobacterium* and introduced into soybean by *Agrobacterium*-mediated gene transfer. As we have described, *Agrobacterium*-mediated DNA transfer resulted in foreign DNA integration, and the chimeric EPSP synthase gene became a permanent part of the soybean genome. These transgenic plants are now resistant to glyphosate because the foreign (but still plant) gene is actively transcribed and translated, and its protein product is an EPSP synthase variant tolerant to the herbicide. The reactions of normal and transgenic plants to Roundup® are diagrammed and summarized in Figure 5.1. Monsanto's glyphosate-resistant soybean was the first transgenic crop plant cultivated for commercialization on a very large scale.

Let us now turn our attention to Liberty®. One of its chemical names is *glufosinate*. Though the names sound somewhat similar, its formula is very different from glyphosate, and so is its mode of action on plants. Plants use nitrate as a source of nitrogen to manufacture their amino acids. Nitrate is present in fertilizers, natural or artificial. When nitrate is absorbed from the soil by plants, cellular enzymes convert it to ammonia. Ammonia is then used by a plant enzyme called *glutamine synthase* to manufacture the essential amino acid, glutamine. Glufosinate is a strong inhibitor of glutamine synthase. What happens when the functioning of

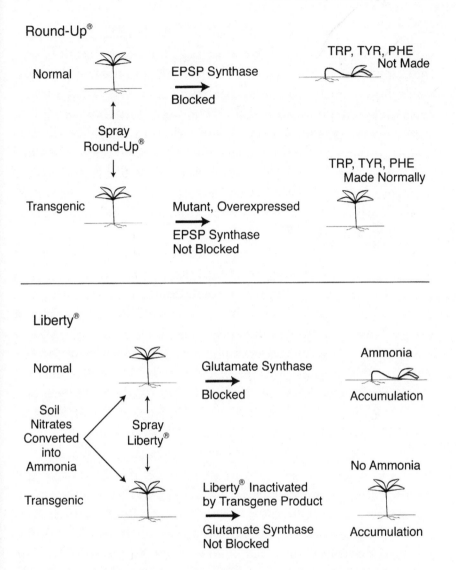

FIGURE 5.1   How Genetically Engineered Plants Become Resistant to Herbicides. Case 1: Roundup®. Roundup® inhibits the plant enzyme EPSP synthase and thereby prevents the synthesis of the three essential amino acids tryptophan (TRP), tyrosine (TYR), and phenylalanine (PHE). In their absence, protein synthesis cannot occur and the plants sprayed with Roundup® die. Case 2: Liberty®. Liberty® kills plants by blocking the plant enzyme gluta-mate synthase.

this enzyme is blocked? The plant cells will continue to import nitrate from the soil and this nitrate will continue to be converted into ammonia. The ammonia will accumulate in the cells because it can no longer be converted into glutamine, the enzyme being blocked by the herbicide. When the ammonia reaches high intra-cellular concentrations, it is very deleterious to all cellular mechanisms. Ammonia is a strong alkali, and its accumulation raises the pH (a measure of acidity or alkalinity) of the cells to lethal levels. The cells die. Glufosinate is safe for humans because we *do not* convert nitrate into ammonia and do not use ammonia to make the amino acid glutamine.

How does one engineer plants for resistance to glufosinate? The approach used was one that introduced into the plants a gene coding for an enzyme that destroys the activity of glufosinate. What is the source of that gene? It turns out that glufosinate is derived from a natural compound, bialaphos, produced by a soil microorganism, *Streptomyces hygroscopicus*. In order to protect itself from its own poison, this bacterium has evolved a defense mechanism that inactivates bialaphos and its glufosinate derivative. This mechanism is simply an enzyme coded for by the bacterium's own *bar* gene (for Basta®-resistance; Basta®, like Liberty®, is one of the commercial names of glufosinate). The function of the *bar* gene (or its variant, the *pat* gene, found in another *Streptomyces* species) product is to tag a chemical group (an acetyl group) onto glufosinate, thereby rendering it inactive. This is because acetylated glufosinate no longer binds to gluta-mine synthase and hence can no longer block its activity. The *bar* and *pat* genes, equipped with an appropriate plant-expressible promoter, were also introduced into plants using the *Agrobacterium*-mediated gene transfer technology as described earlier. To summarize, plants engineered with a bacterial *bar* or *pat* gene take up the herbicide glufosinate, as all sprayed plants do, but instead of dying from ammonia poisoning, they chemical-ly alter glufosinate and inactivate it. These plants then survive because ammonia does not accumulate. The reactions of normal and transgenic plants to Liberty® are diagrammed in Figure 5.1. An important note: The protein products of the *bar* and *pat* genes are destroyed within seconds after exposure to human digestive gastric fluid, precluding any allergic or toxic reaction after human ingestion. The *bar* gene technology was first tested under field conditions with transgenic tobacco and potato plants by the Belgian company Plant Genetic Systems (PGS), now part of AgrEvo. Interestingly, one of the founders of PGS is Marc Van Montagu, one of the two heads of the University of Ghent team that did so much to elucidate the crown gall biological processes. Glufosinate-resistant canola plants have been grown for commercial purposes for several years, especially in Canada.

Interestingly, Monsanto has also isolated a gene that codes for the inactivation of Roundup®, either by chemical modification, as in the case of Liberty®, or by some other means. Details, however, are difficult to determine because the company guards them as proprietary information.

Genes that determine resistance to other herbicides have been identified in and cloned from several other bacterial or plant species. Examples include resistance to 2,4-D, chlorsulfuron, Bromoxynil®, and Dalapon®. The modes of action of these herbicides are completely different from those of Roundup® and Liberty® and sometimes are poorly known. Experimental plants have been engineered with these genes, and they do resist herbicide action. However, no commercially available crop plants engineered with these genes yet exist. Resistance to 2,4-D, Bromoxynil®, and Dalapon®, as in the case of Liberty®-resistance, involves genes whose enzyme products chemically modify the herbicide and make it inactive in plants. This means that sprayed resistant plants will to some extent accumulate the chemically inactivated herbicide, raising the question of human vulnerability to the inactivated herbicide. Indeed, even though a particular herbicide may have very low human toxicity, it does not necessarily follow that its enzymatically inactivated derivative will maintain this low human toxicity. Therefore, extensive testing of these transgenic plants is necessary before considering commercialization of products that would affect humans.

Weeds are not the only pests that affect crop plant yields. Many types of insects feed on crop plants and wreak havoc on harvests. Two well-known examples of plants that are susceptible to insect attack are corn and cotton. Some four-winged insects, such as the European corn borer and the cotton bollworm, burrow into these plants and feed on their tissues. The results are, of course, very sick plants. Other plants are also devastated routinely by insects, but so far only corn and cotton have been genetically engineered to resist attack.

The process used to achieve this was very simple. A bacterial toxin, called Bt toxin, had been used by farmers, including many organic farmers, for more than thirty years to obtain protection against insect predation. This toxin, in fact a natural insecticide, is used in the form of a spray. Major problems associated with this practice are the high cost of production of the toxin and its quick inactivation in the field. Why not generate plants transgenic for the gene coding for this toxin? In theory, they would be insect-resistant because they would make the Bt toxin themselves, and all of the toxin would be available to exercise its insecticide action instead of degrading in the soil. This is exactly what was done. The Bt toxin gene was cloned from its host, the bacterium *Bacillus thuringiensis*, engineered for expression, and transferred into plants by *Agrobacterium*-mediated gene transfer.

This toxin is insect-specific, but how does it work? Bt toxins (there is a whole range of them) are proteins made by the same bacterial species mentioned earlier. When ingested by insects, these proteins are partially degraded in their midgut, releasing fragments that are toxic to their intestinal cells. These protein fragments are toxic enough to disrupt the insects' digestive physiology and kill them.

Humans are not sensitive to Bt toxins because the human digestive tract is fundamentally different from an insect's. The insect's midgut environment is alkaline (high pH), so that Bt toxins can exert their effect. Human gastric fluid, on the other hand, is highly acidic (very low pH) and leads to Bt toxin inactivation within seconds after ingestion. Furthermore, if anything, the safety of Bt toxin in agriculture has been demonstrated by default. As stated earlier, some organic farmers use Bt toxin (although they don't necessarily tell their consumers about that), and it has been used abundantly for decades by nonorganic farmers. Had Bt toxin been toxic to humans, we would all be dead or severely crippled by now. It is therefore reasonable to think that corn engineered with a Bt toxin gene is safe to eat. Unfortunately, the term *toxin* immediately elicits a negative reaction. It is too late to change this terminology, but let us remember this analogy: A simple aspirin tablet can kill a cat. Therefore, if cats could, they would call aspirin a *toxin*. These semantics are not sufficient reason for humans to stop taking aspirin. The same holds true for Bt toxin; it kills certain insects but is without effect on humans. Numerous tests conducted with Bt toxin gene-engineered corn confirm this fact.

Transgenic cotton was also engineered with Bt toxin genes, using the technology first applied to corn. Transgenic corn and cotton are both sold by Monsanto and are widely grown in the United States. It should be noted that in the case of corn, the European corn borer is only one of a number of insect species that prey upon this plant. The Bt toxin gene used to protect corn does not necessarily kill other insects, meaning that spraying with pesticides is still often required for full protection against insect damage.

What about viruses? They too can have negative effects on crop production. The traditional way to avert or reduce the impact of viral infections is to introduce resistance genes from a wild gene pool (present in the wild relatives of crop plants that grow in nature) into cultivated varieties through conventional breeding. This technique is extensively used in lentils, for example. There are, however, two drawbacks to this approach. First, wild varieties, even virus-resistant ones, do not possess qualities found in their domesticated counterparts (for example, yield and nutritional value). Therefore, the undesirable wild traits have to be bred out of the hybrids by repeated crosses with the cultivated species. This can take a long time—often several years. Second, viruses mutate and can, over time, defeat the natural resistance imparted by a wild plant gene.

Several new strategies based on genetic engineering have allowed production of virus-resistant plants. A certain type of virus-resistant squash, developed from this technology, has reached the marketplace. What follows is an explanation of how virus resistance or tolerance can be induced using plant transformation techniques. All viruses have a genome made of either DNA or RNA. The nucleic acid (RNA or DNA) is contained in a coat made of one or several proteins. When viral particles invade a plant cell, the coat is stripped away from the nucleic acid, which can then be replicated and expressed to make more viral particles. It was reasoned that preventing the "uncoating" of the viral nucleic acid would also prevent its multiplication and expression, because the enzymes necessary for these functions would not have physical access to DNA or RNA covered with a coat protein. This reasoning proved to be correct; plants engineered with *Agrobacterium* T-DNA containing a viral gene coding for coat protein production are resistant to virus infection. Once the coat protein gene is introduced, it constantly produces coat protein molecules that re-envelop any invading viral nucleic acid.

This technique is just one of several that allow production of virus-resistant transgenic plants; the strategy that is used depends on the type of virus considered. One pioneer in this area is Roger Beachy of Washington University, St. Louis. In addition, the French have used a similar approach to protect grapevines from viral attacks. It is still unclear whether genetically engineered French wine produced from these grapes will ever please our taste buds. In another instance, papaya growers in Hawaii have warmly embraced the transgenic technique to produce fruits unblemished by ring spot virus infection and have reported that consumer response has been very positive.

We have also used the virus coat protein technology in my laboratory at Washington State University to generate transgenic pea plants resistant to a certain virus, pea enation mosaic virus (PEMV). These plants are for research only and are not meant for human consumption, however. Like many researchers before us, we noticed that the expression of the virus coat protein gene varies in individual transgenic plants. In other words, some of our engineered pea plants showed high tolerance to the virus, whereas others fared barely better than nonengineered plants. This situation is often encountered in transgenic plants; introduced transgenes are not expressed to the same level in different transgenic individuals. The reasons for this are not completely clear; variations in transgene expression may depend on the point of integration of the transgene in the host genome, or they may depend on the number of copies of the transgene that were transferred to the recipient plants. Therefore, genetic engineers must routinely produce many transgenic individuals and select those that express the new gene to its fullest extent.

Let us now consider one last class of biological agents that can cause injury to plants: microbes. Plants can be damaged by fungal and bacterial attack. Although basic research on these problems is making great advances, no resistant commercial transgenic plants have been produced yet. There is no single way to make crop plants insensitive or less sensitive to all pathogenic bacteria and fungi. Each crop species can be attacked by a specific set of microbes, against which no single universal defense mechanism has been discovered. In fact, the virulence of a particular pathogen for a particular plant variety depends on the genetic makeup of *both* the pathogen *and* the host. Wild relatives of domesticated plant species sometimes harbor resistance genes to some pathogens. In many cases, these genes have been introduced into cultivated varieties by conventional breeding. In addition, the use of genetic engineering techniques has demonstrated that resistance to microbial pathogens can be achieved by genetic transformation. One example is the production of tomato plants resistant to a particular variety of *Pseudomonas syringae*, the bacterial agent that causes bacterial speck in tomatoes. The Rio Grande tomato variety is naturally resistant to *P. syringae*. A piece of DNA cloned from Rio Grande was identified as the carrier of a potential resistance gene. When this piece of DNA was transformed into the sensitive tomato variety, called *Moneymaker*, by *Agrobacterium*-mediated gene transfer, it made Moneymaker resistant to *P. syringae*. Base sequence analysis of the resistance gene revealed that it codes for a kind of enzyme called a protein kinase. These kinds of enzymes are often used by cells in what is called a *signal transduction pathway*. This signal transduction pathway can be thought of as steps that activate plant defense mechanisms. When this series of steps is absent, as in the Moneymaker tomato, the plant does not activate its defenses and is thus susceptible to infection by the pathogen. Several pathogen-resistance genes have been identified and cloned from a variety of plant species but have not yet been used commercially. Much pioneering work of this nature was done by Steven Tanksley's laboratory at Cornell University. Finally, researchers at the University of Louvain, Belgium, have succeeded in transforming banana plants for resistance to the most serious fungal disease that preys on that plant. These bananas are not yet on the market.

It should be pointed out that engineering plants using plant genes that confer resistance to bacterial or fungal pathogens is as close as one can get to conventional plant breeding. The big difference, however, is that the genetic engineering approach is much faster and much more specific. As described earlier, conventional breeding mixes *all* the traits of two plant varieties, some of which then have to be bred out because they may not be commercially desirable. This process takes years, as opposed to genetic engineering techniques that take only a few months. Further, breeding is impossible with plant species that are triploid, such as the banana. Gamete production always involves even numbers of chromosomes, not odd numbers, as in triploid bananas. These plants are propagated by the cultivation of cuttings because they are sterile. In such cases, only genetic engineering techniques allow the introduction of new genes and traits.

We have seen in this section that much effort is devoted to protecting crop plants from chemical and biological injury. The ultimate goal of this type of research is to increase plant productivity and to decrease waste. We will see next that productivity (yield) can also be enhanced by genetically manipulating the *metabolism* of plants.

Boosting Yields Through Enhanced Photosynthesis

Increasing the yield of crop plants is probably as old an objective as agriculture itself. Classical breeding techniques, based on Mendel's laws, have been extremely successful in producing high-yield plants like wheat, barley, rice, and corn. To achieve this result, breeders make crosses between different varieties that belong to the same species or that are closely related (sexually compatible). Then, often after years of work, they select the most desirable offspring of these crosses and release them as new varieties. The difficulty with a trait such as grain production is that it is not under the control of a single Mendelian gene. Rather, many genes are involved, and this multiplicity is what makes the production of high-yielding varieties so time-consuming and arduous. Further, gene transfer technology that deals with multiple genes is still in its infancy. For this reason, the transfer of yield-determining genes has not yet been achieved through genetic engineering. To complicate matters further, it may well be that the current plant gene pool has been mined as much as it can be. In other words, it may be that classical breeding is approaching its limits because all the high-yield genes available have already been used for breeding purposes. The gene pool may well have been exhausted. What, then, can be done to improve yields further?

Maurice Ku, a researcher at Washington State University, in collaboration with his Japanese colleagues, used an entirely new approach to tackle this problem—with rice. They based their reasoning on the fact that biochemical reactions of photosynthesis that convert atmospheric carbon dioxide into sugars are inefficient in rice plants and potentially could be improved. Indeed, rice is part of a group of plants (called C3 plants) that absorb carbon dioxide from the air much less efficiently and use more energy than most other plants (called C4 plants, corn being one example) to achieve the same result, the conversion of carbon dioxide into sugars mediated by sunlight. Ku and his colleagues hypothesized that transferring key genes from a C4 plant into rice could potentially enhance carbon dioxide fixation. Interestingly, rice possesses these genes but they are silent. Ku's group cloned three genes from corn, coding for the three C4 enzymes, NADP-malic enzyme, phosphoenolpyruvate carboxylase and pyruvate, orthophosphate dikinase, and transferred them, one at a time, into rice, using a high-level expression vector. Normally, one would expect that addition of a single gene would not drastically enhance carbon dioxide fixation. Yet, this is what happened: All three transgenic lines showed enhanced carbon dioxide utilization and a 20 percent increase in rates of photosynthesis. What is more, the yield of these lines, in terms of grain production, was increased by 10 percent to 35 percent (as shown in Figure 5.2). The precise reasons why single genes have such an important effect on these transgenic plants are still under investigation, but this discovery bodes very well for the productivity of this all-important crop. High-yielding transgenic rice is not yet available to the public, however.

FIGURE 5.2   Control (top) and Transgenic (bottom) Rice Plants of the Kitaake (*japonica*) Variety. The higher number of rice grains on the engineered plant is quite evident (courtesy of Dr. Maurice Ku, Washington State University).

Produce Quality and Nutrition

My students do not know the real taste and texture of a tomato. Unless they grow their own, they believe that a tomato is a red sphere with no taste and hard enough to withstand a few sets of tennis. This has nothing to do with biotechnology, it is a consequence of agricultural practices that dictate that tomatoes must be harvested when green and hard in order to avoid bruising and rotting on the shelf. These inedible tomatoes must be blasted with ethylene gas to trigger genes that impart red color. This produces the normal appearance of tomatoes as used in salads and spaghetti sauce. This treatment improves color but does not change the bland taste nor the hardness of the immature fruit. Ethylene treatment is in itself inoffensive; this is the same gas that is produced by the tomato fruit itself during the natural process of ripening. Biotechnologists thought that it would be interesting to develop bruise-resistant tomato fruits that would ripen on the vine, thereby gaining their red color naturally together with some flavor.

This project led to the very first commercial genetically engineered plant product available in the United States, the Calgene FlavrSavr® tomato that appeared on some grocery store shelves in the late 1980s. This engineered tomato was a market flop, and Calgene no longer exists as an independent company. FlavrSavr® holds the dubious honor of having started the protest against GMOs used in human nutrition. We will see the reasons for this protest later. But first, how do tomatoes soften during ripening? This is the process that should be slowed down, to avoid bruising, while allowing the other aspects of ripening, color and flavor, to develop and to proceed normally. Tissue softening is due in large part to the production of a natural enzyme, polygalacturonase, in the mature fruit. The function of this enzyme is to degrade tough cellular structures, until the tomato turns liquid and its seeds can be released. Inhibiting polygalacturonase should retard this process and keep the fruit firm. This is exactly what the Calgene scientists did. For this purpose, they used what is called the *anti-sense* technology[1] based on the use of the natural tomato gene that codes for polygalacturonase. It worked. The FlavrSavr® remained firm while reddening on the vine and acquiring a bit more flavor than other commercial tomatoes. Yet, public perception of GMOs quickly grew so negative that FlavrSavr® was left to rot on the shelves. It is no longer available.

Bluntly stated, there are currently no commercially available transgenic crop plants that possess improved qualities that appeal to the consumer. Indeed, insect- and herbicide-resistant crop plants benefit biotechnology companies and growers, but they do not really benefit people who shop for food. Practically all projects that *could* benefit the consumer are bogged down in regulatory battles hinging largely on public mistrust of transgenic technologies. This even applies to enriched golden rice. As we know, the natural color of rice grains is pearly white. A Swiss-German team has succeeded in turning rice grains golden by genetically engineering the plant for vitamin A production in its seeds, the rice grains.

This project has nothing to do with color itself; the goal was to create rice rich in vitamin A, or, to be exact, provitamin A, beta-carotene, the compound abundantly present in carrots. Vitamin A deficiency in children is chronic in Southeast Asia and affects up to 70 percent of children aged one to four years. One of the many consequences is blindness. It is estimated that more than 120 million children are at risk worldwide, and that 1 to 2 million deaths annually could be prevented if vitamin A were administered to these children. Rice is a staple food in much of the world, but unfortunately, it contains no vitamin A in its polished grains. A normal daily diet of vitamin A-containing rice would supply enough of the vitamin to eradicate deficiency and many health problems in at-risk children.

The Swiss-German team[2] created just this kind of rice. Here is how they did it. Since provitamin A is processed by the human body into vitamin A, it is only necessary to engineer rice plants that produce provitamin A. Rice grains make a compound located at the beginning of the provitamin A synthetic pathway but lack the enzymes necessary to convert this compound into provitamin A itself. Rice plants were engineered with two genes that code for the enzymes necessary to achieve the conversion. One gene comes from the daffodil and the other comes from a bacterium found in the soil, *Erwinia uredovora*. All genes were engineered with promoters ensuring expression in the grains, and rice plants were transformed with these chimeric genes yet again by *Agrobacterium*-mediated gene transfer. The results of the experiments were rice plants that produce golden grains, the natural color of provitamin A. The amount of provitamin A produced by this transgenic rice is sufficient to provide all the necessary vitamin A in a regular daily diet of rice. Also, since provitamin A has very low toxicity (as opposed to vitamin A itself), overdosing will not be a problem. As I see it, golden rice should have a bright future.

One last example regarding produce quality is worth discussing, because it blends traditional plant breeding and biotechnology. It has been known for decades that hybrids in both the plant and animal worlds show more vigor, meaning better resistance to disease, higher yield, and better performance in a variety of environmental conditions. The production of hybrid seeds is not simple, because many crops are self-fertilizing, like Mendel's peas. Even if two different parents are planted close together, most fertilization events will take place between gametes located in a single parent. Cross-fertilization will not occur spontaneously. We have seen that Mendel was able to cross-fertilize his pea plants by manually emasculating one parent and manually fertilizing its pistil with pollen harvested from the other parent. Now, doing this type of manipulation on an industrial scale is unimaginable. Just think about a field planted with 10,000 tomato plants, each carrying many flowers, every single one of which would have to be castrated and manually fertilized with pollen from another parent. We have yet to create a machine that could perform this function on the grand scale necessary for commercial farming.

In some cases, spontaneous mutations took care of this problem by making some individuals male-sterile, though keeping them female-fertile. In these plants, pollen grains are nonfunctional and unable to fertilize ovules. Self-fertilization is impossible. Thus, these plants can only be propagated by fertilization with pollen from another male-fertile plant. The product of this fertilization is, of course, a hybrid. A well-known example of male sterility is found in corn. Unfortunately, examples of male-sterile crop plants are few and far between, making hybrid production a real challenge. Plant biotechnology came to the rescue, at least on paper, by manipulating pollen grain formation. A Belgian-U.S. team[3] reasoned that if anthers (the male sexual organs of flowers) were rendered unable to produce pollen grains, the plant so engineered would be male- sterile. The team achieved this feat by expressing an enzyme that degrades RNA in the oilseed rape plant. Recall that a DNA gene can only be expressed if its RNA copy is made first. Thus, the researchers transferred a ribonuclease gene (coding for the RNA-degrading enzyme) from the bacterium *Bacillus amyloliquefaciens* into these plants. This gene was engineered for expression in anther cells only. The result was that all RNA copies of all the DNA genes were destroyed in these anther cells, from which pollen grains normally develop. As a result, no pollen grains were made, but since the ribonuclease was only expressed in anther cells, the rest of the plant was normal, including its female sex organ. These engineered plants were fertilizable by pollen from normal, nonengineered plants and gave rise to progeny seeds. Male sterility had been achieved by genetic engineering through a methodology applicable, in principle, to all crop plants. This approach should greatly facilitate high-quality hybrid seed production in crop plants. However, to my knowledge, this technique has not yet been applied, except in the laboratory and in some field trials and in spite of the fact that successful results were obtained more than ten years ago.

Thinking about the enormous range of potential plant biotechnology applications, it is perhaps surprising that so few genetically engineered crop plants are presently commercialized. Undoubtedly, this is due in part to a negative view of biotechnology, but, in some cases, this phenomenon is due to the stranglehold of industrial patents on free product dissemination. This will be discussed next.

Patent Protection

Transgenic plants currently in production (as with corn, soybean, and canola) were developed by corporate scientists. Nonetheless, the *basic* discoveries described earlier, without which plant biotechnology would be impossible, were mostly made in university laboratories. Academic scientists have been remarkably incompetent at protecting their intellectual property. Although some have had the foresight to protect their work, by and large, they never realized that their work would have the impact—good or bad—that it now has on society. The field was wide open for corporations to claim ownership of basic technologies that most of them hardly helped develop (except through financial support). The result has been patent wars between companies, which is perhaps a good thing since the battles have slowed down the eagerness to market dubious applications of biotechnology.

One of these possible applications is the infamous *terminator technology*. In this scenario, crop plants engineered for desirable traits would be equipped with a lethal gene whose activity would be triggered by a compound present in a regular spraying solution, sold by the same company that made the engineered crop plants. The result of this spraying would be seeds with normal nutritional qualities, but that would be *unable to germinate*. The benefits to the seed company would be obvious: The farmers could still sell their high quality grain but would be unable to keep part of their harvest for further planting because these seeds would be dead ends, unable to grow. Such practice would ensure the company's total power and monopoly over crop growers. Rumors (true or false) that Monsanto was going to apply this terminator technology to its presently available engineered corn, soybean, and cotton spread like wildfire in the scientific and farming communities. The resultant protest required that Monsanto refute the rumor as being just that, a rumor. Nevertheless, I believe that vigilance is necessary. After all, there are stringent antitrust laws in the United States, and biotechnology companies cannot be excepted.

Ironically, the terminator technology was resurrected in the year 2000 by a somewhat unexpected actor: the U.S. Department of Agriculture (USDA). This federal agency holds patents that it now intends to commercialize. The consequences of this decision are not yet known. Finally, it is comforting to note that Monsanto has recently decided to provide royalty-free licenses to accelerate the development of golden rice varieties. However, Monsanto can only lay claim (and grant permission) to a small part of the technology that led to the creation of provitamin A-rich rice. It is estimated that between twenty-five and seventy other patented techniques and materials were used in the production of golden rice, an indication of the extreme complexity of the problem. There are now plans to sell golden rice to developed countries that, in turn, would give it away for free to poor nations. We will have to wait and see what materializes.

The creation of golden rice should be seen as a fantastic success, perhaps even worth a Nobel Prize. Unfortunately, one of the benefactors of this research, the Rockefeller Foundation, pulled the plug on this project early in 2000, possibly due to the negative press accorded plant biotechnology, but also perhaps because the Foundation considered that the results of the project had been achieved. At the time of this writing, it is unclear when or whether golden rice will benefit the millions of children at risk for blindness and other serious ailments.

Last but not least, it is encouraging to note that Ingo Potrykus, one of the leaders of the golden rice project, was featured on the July 31, 2000, cover of *Time* magazine, a well-deserved honor and excellent publicity for his worthwhile contribution to human welfare.

6

Other Applications of Plant Biotechnology: Applications for Plant Resistance to Environmental Stress, for Human Health, and for Protection of the Environment

M Y FRIEND AND COLLEAGUE, geneticist László Márton, has had an eventful life. Born and educated in Hungary, he planned first to be a forester and earned a college degree in this field. That interest did not last for long, however. In spite of his love of nature, which included fishing for carp and hunting for bear (some say he doesn't even need a gun to get a bear; he is so large and strong that he could just wrestle one), the call of genetics became irresistible. He eventually earned a Ph.D. in plant genetics and started working for the Hungarian Academy of Sciences. I first met him there in 1976, in a country that was then under communist rule.

Márton was not a party apparatchik, however. This became apparent one day when we were doing an experiment in his lab, as we were nervously looking for some kind of support to prop up a rack of test tubes in the right position. Nothing seemed to fit until his gaze rested on a volume of Lenin's thoughts, which was mandatory in Hungarian labs back then. Beautiful! The book was just the right thickness for what we needed. With a sarcastic grin, Márton uttered, "This is the best use I've ever had for this book."

Márton left Hungary a few years after this episode, to spend extended periods of time doing research in the Netherlands, Belgium, and the United States. During that time, he made major contributions to plant genetic engineering. For over a decade, he has been a professor at the University of South Carolina in Columbia and the director of their plant biotechnology center. One of Márton's current projects is the genetic engineering of aquatic plants for the environmental remediation of heavy metal pollutants in South Carolina's coastal salt marshes. This example illustrates that plant biotechnology extends far beyond its applications to genetically modified food products.

We saw in the previous chapter examples of genetically modified plants that are either in production (GM soybean, cotton, corn, canola, and papaya) or possibly on the verge of being grown on a large scale (GM rice). This chapter deals with potential applications that are still at the experimental stages and for which mostly laboratory plants (such as *Arabidopsis*), not necessarily food plants, are used as experimental material.

Salt and Frost Tolerance
In addition to biological stress (viruses, bacteria, fungi, weeds), plants can also be subjected to physical stress. Here I cover two examples: soil salinity and frost, and what biotechnology can do to increase plant tolerance to these agents. Salinity of soils is a serious factor that limits productivity of agricultural crops. High concentrations of sodium have deleterious effects on metabolic processes and water balance in plants. Further, high salinity causes loss of arable land and is a direct cause of deforestation, which in turn leads to greater salt accumulation. It would be of great value to grow crop plants in high-salinity soils in order to break

this vicious cycle. Some plants are naturally salt-tolerant, and basic research has shown that this is due to their efficient ability to accumulate externally supplied sodium into the vacuoles of their cells. All plant cells contain vacuoles; these are cellular bodies used as storage compartments for some proteins, pigments, and waste products. They also regulate the amount of salt present inside plant cells. Salt-sensitive plants also possess vacuoles that regulate intracellular sodium concentration, but their salt uptake mechanism is less efficient than in salt-tolerant plants.

In 1999, a Canadian research group reasoned that salt-sensitive plants conceivably could be rendered salt-tolerant if a gene responsible for sodium transfer into vacuoles were overexpressed. Remember that overexpression of a gene consists in cloning the coding sequence of this gene behind an active promoter, as was done in the case of Roundup® resistance. The Canadian group isolated a gene called *AtNHX1* from the model plant *Arabidopsis*. This gene is responsible for sodium transport into vacuoles. They cloned the coding sequence of this gene under the control of a hyperactive promoter and introduced the chimeric gene into normal, salt-sensitive *Arabidopsis*. Sure enough, transgenic *Arabidopsis* became sodium-tolerant.

This technique could be applied readily to crop plants, but there was a problem: Transgenic salt-tolerant *Arabidopsis* obviously accumulates sodium in its vacuoles and, presumably, transgenic crops would do the same. Since sodium is sequestered inside vacuoles, the plants are protected, but what about people who might eat vegetables engineered in such a way? Certainly, their dietary sodium would increase, not always a good thing. Thus, in the case of salt-tolerance, agricultural yields could be boosted by the use of transgenic techniques, but the possible consequences on human health must be evaluated carefully before widespread applications are considered. One way to alleviate medical problems would be to restrict expression of the salt-tolerance gene to the nonedible parts of the crop plant. This has not yet been accomplished, and there are presently no transgenic salt-tolerant plants available on the consumer market.

Another problem associated with soil quality is acidification due to acid rain. Acid soils dissolve the element aluminum, which is extremely toxic to plants—it greatly inhibits root development. It is estimated that about 40 percent of the arable land worldwide is at risk for aluminum toxicity. It is known from laboratory experiments that certain natural organic acids can alleviate aluminum toxicity by forming a complex with its salts. This complex is not absorbed by plant roots, which are therefore not inhibited, and so allows normal plant growth. A Mexican group has genetically engineered papaya plants for resistance to aluminum. They cloned a gene that codes for citrate synthase (an enzyme that catalyzes the formation of citric acid, as found in citrus fruits) from the soil bacterium *Pseudomonas aeruginosa* and recloned it under the control of a strong plant promoter. The chimeric gene was introduced into papaya cells using the biolistics technique. Transgenic papayas were shown to overexpress the foreign gene and, consequently, overproduce the natural plant compound citric acid, an organic acid able to neutralize aluminum salts. Interestingly, significant amounts of citric acid were excreted into the plant growth medium by the transgenic plants, showing that aluminum became neutralized by citric acid outside the roots and was thus not absorbed. Therefore, these transgenic plants did not accumulate aluminum. This result is in contrast to the approach just discussed, in which excess sodium was sequestered inside the plant cells and would eventually be ingested by the human consumer. Still, there is no evidence yet that aluminum-resistant plants are being commercially produced.

Low temperature, another physical type of stress, also can dramatically reduce crop plant yields even in areas where freezing is rare, as Florida citrus fruit growers well know. A much publicized application of transgenic techniques is the possibility of engineering plants with a fish antifreeze protein to increase frost tolerance. Flounders are fish that thrive in marine environments at subfreezing temperatures. Flounders (and other fish like them) do not freeze solid in the Arctic ocean in part because they produce a special protein that lowers the freezing point of the water within their tissues. The gene coding for this protein has been cloned and transferred to laboratory plants. And, yes, this fish gene affords some protection against frost. This idea seems particularly repulsive to many people because it mixes the genes of vegetable and fish into one single organism. This is of course an exaggeration; plants containing a single fish gene and hence a single type of fish protein do not become fish. Nor does transgenic corn containing a single bacterial gene suddenly become bacterial. Nevertheless, some people have an aversion to fish (I am one of them), and strict vegetarians would consider plant material engineered with a fish or other animal gene to be unacceptable. Moreover, some people are allergic to fish (although the antifreeze protein may not be an allergen all by itself).

I do not believe that crop plants expressing this fish gene are anywhere close to commercialization and will probably never reach that stage. This is obviously due in part to public concern and the fact that the fish gene is unnecessary for protection against frost. Interestingly, plants also contain genes that respond to cold stress and can provide protection against mild frost on their own. This being the case, one could engineer plants to over-express their own genes and simply magnify their innate ability to withstand low temperatures.

Freezing tolerance in plants, called *cold acclimation*, can be increased by gradual exposure to nonfreezing temperatures below 40°F. Some cold-responsive genes are induced, or turned on, during the acclimation period. Several classes of genes are induced by cold temperatures and were given names such as *cor* (for *co*ld *r*esistance) and *eskimo*. The function of some of these genes seems to be to stabilize cellular membranes by arranging their lipids (fats) around proteins in a particular fashion. Other genes involved in freezing tolerance do so by way of an as-yet-unknown mechanism.

Much research remains to be done in order to understand how cold resistance genes work in plants, but the important point is that they *do* exist and could be manipulated for enhanced expression, thus bypassing the use of any fish or other nonplant genes. Preliminary results obtained with overexpressed freezing tolerance plant genes are encouraging. Much basic work on freezing tolerance genes is being done in the laboratory of Michael Thomashow at Michigan State University. In conclusion, putting a cold-resistant fish gene into plants had an enormously negative psychological impact on the public and was a totally unnecessary experiment. I predict that plants engineered with this fish gene, if they still exist, will remain laboratory curiosities.

Let us now turn to one of the most ridiculous adventures of the new genetic technology, one that involves frost resistance. About twenty-five years ago, Steven Lindow of the University of California at Berkeley started thinking about the process of ice formation on plants. Freezing water kills plants (and other living things) by forming crystals inside cells. These crystals, when they form, shear delicate cellular structures and damage them irreparably. Also, water freezes more easily when ice-crystal formation is facilitated by any kind of structure that causes nucleation, that is, crystallization of water molecules into ice. Lindow figured out that bacteria living naturally on the surface of plants served exactly like the kind of structure that facilitates ice nucleation. What is more, he identified a particular bacterial surface protein that triggered ice-crystal formation as the temperature fell below a certain point. He then engineered these bacteria in such a way that they would no longer produce this surface protein and would then possibly prevent ice nucleation if sprayed on plants. His reasoning was right—the bacteria deficient in this protein no longer induced ice-crystal formation on the treated plants and protected them from frost injury.

Lindow would face profound challenges, however. The experimental plots established to test his idea in the field were repeatedly savaged by activists who opposed genetic engineering. These actions were particularly absurd because the bacteria engineered by Lindow had in fact *lost* a gene, the ice nucleation gene. No monstrous organism was ever produced in these experiments. I met Lindow at the time his field experiments were sabotaged. He had come to the University of California–Davis campus, where I was working at the time, to find intellectual and moral support for his idea. He looked to me like a hard-working scientist, not like some vicious Dr. Demento ready to blow up the world. The unbelievable irony is that the bacteria he modified to prevent ice formation in plants are used today in large amounts in their normal configuration (that is, containing the ice nucleation gene) to generate snow on ski slopes during warm winter weather. I wonder sometimes if the people who destroyed Lindow's test plots strapped on their skis right after their acts of vandalism and got an additional adrenaline rush by barreling down slopes made with basically the same bacteria he had domesticated to reach the opposite result. Lindow's crippled bacteria have yet to be accepted in engineering frost resistance in agricultural plants.

Plant Biotechnology and Human Health

Projects aimed at improving the nutritional values of crop plants, other than golden rice, are still in development. These include manipulating plants for micronutrient and fat content. Micronutrients are, by definition, compounds present in small amounts in foods that are needed for good health. Examples include vitamins, iron, iodine, and selenium. It is estimated that 2 billion people are at risk for iron deficiency and 1.5 billion are at risk for iodine deficiency. Most of these people live in developing nations, but even those living in developed countries are not risk-free, given the much greater importance now given to a diet rich in vegetables. Not all edible plants contain sufficient amounts of all micronutrients, and their intake depends on the nature of the human diet. Staple foods such as wheat, corn, and rice are examples of plants in which many essential micronutrients are not present in adequate amounts.

We have seen what was achieved in the case of golden rice. Similar progress has been made with vitamin E. Genetically engineered *Arabidopsis* has been coaxed to produce nine times more vitamin E in its seed oil through overexpression of one of its own genes, the gamma-tocopherol methyltransferase gene. Similarly, genes involved in iron uptake in plants have been manipulated to boost iron retention by plant cells. All these observations and discoveries show great promise. In this research, we must be mindful of producing edible plants that contain toxic levels of these elements. Too much iron or too much vitamin E, for example, is not a good thing. Thus, humans consuming such engineered plants would have to know exactly how much plant material to ingest in order to reach healthy micronutrient levels. But they would also have to know when to *stop* eating in order to avoid harmful effects. Ironically, eating might become a little like taking medicine and the phrase "Do not exceed the recommended dose" might accompany some engineered vegetables.

As for fat content of our vegetables, the trend now is to manipulate fatty acid synthetic pathways in order to increase the content of unsaturated fatty acids in plant oil. Unsaturated fatty acids are healthier than saturated fatty acids and are also more stable in frying and cooking. Oleic acid, a monounsaturated fatty acid, is seen as particularly desirable for a healthy diet. Transgenic soybean lines, whose oil contains up to 85 percent oleic acid as compared to 25 percent in normal soybeans, have been created by DuPont, the giant chemical company. Here, also, much progress has been made, but to my knowledge, no soybean lines with improved (or is it better to say "modified"?) fatty acid content have been exploited commercially.

Finally, from a purely medical viewpoint, plants are now being engineered for use as a source of vaccines. Vaccination consists in injecting individuals with a dead or disabled pathogen (generally a virus), or sometimes a part of a pathogen (called the *antigen*), in order to trigger an immune response (through the production of antibodies) in the treated person. Some vaccines can be taken orally. When the immunized individual is later exposed to the live pathogen, his or her immune system will recognize the invader and neutralize it. Vaccination has been extremely successful in preventing a cohort of disabling or lethal viral and microbial infections in human populations. This method was how smallpox was eradicated throughout the world several decades ago and how polio has been nearly eradicated at the time of this writing.

However, access to vaccines is still a problem in many parts of the world. There are always the difficulties of storage when refrigeration is needed, and sterile syringes, when required, are not always easy to come by. Plant-based, edible vaccines would solve these problems. This is the approach taken by Charles Arntzen (now at Cornell University) and his team at Texas A&M University. The reasoning is that often, it is not necessary to use the whole pathogen to elicit an immune response, only its toxic component. This is the case for gastrointestinal diseases caused by *Vibrio cholerae*, the causative agent of cholera, and pathogenic *E. coli*. These two bacteria cause harm by secreting protein toxins that injure intestinal cells and cause severe diarrhea. The genes coding for these toxins have been cloned. The idea, then, is that by expressing partial toxin genes in edible plants the treated individual's body would trigger an immune response to a disabled, safe partial protein toxin. A partial toxin gene was cloned under the control of a plant-expressible promoter and introduced into potato plants by (yes, *again*) *Agrobacterium*-mediated gene transfer. Experiments demonstrated that the partial protein toxin was produced in potato tubers, which were then fed orally to mice. The mice started producing antibodies to the toxin. They had thus been immunized by consuming potato tubers made trans-genic for the toxin gene.

Further refinements are needed, however. Observed immunity levels were not quite as high as is desirable, and even though mice do not mind consuming raw potatoes, the same cannot be said of humans. Since boiling or french-frying the potatoes would destroy the immunological properties of the toxin, another plant host that can be eaten raw must be found. The banana may be a good candidate.

Nonetheless, in subsequent tests, human volunteers were fed raw potato tubers engineered to produce an immune response against pathogenic *E. coli*. The results of these tests were very encouraging. In July 2000, the Cornell University group, in col-laboration with clinicians at the University of Maryland, success-fully immunized nineteen out of twenty (95 percent) volunteers against the Norwalk virus, a leading cause of food-borne disease around the world. In this case, too, potato tubers were engi-neered with a viral component gene and consumed raw.

Since the technical term for toxins that elicit an immune response is *antigens*, the toxins produced in transgenic plants can be dubbed *plantigens*. Further, since antigens trigger an immune response that results in the production of antibodies, one could imagine plants made transgenic with genes coding directly for antibodies. Consuming these plants would provide human patients protection against some infections, much as what is achieved with gamma-globulin injections for partial protection against various infectious diseases. These antibodies could there-fore be called *plantibodies*. Also, since antigens and antibodies are considered pharmaceuticals, the practice of producing them (and perhaps other medicines) in transgenic plants has been called *pharming*. A great advantage of these plantigens and plan-tibodies is that they cannot be contaminated by animal viruses, including human viruses—always a possibility with vaccines pro-duced from animal cells cultivated in vitro and antibodies isolated from human donors. Further, even though plants are hosts to many types of viruses, these do not infect humans. Work on edible vaccines continues and, I think, represents one of the most interesting spin-offs of plant genetic engineering.

How Plant Biotechnology Could Be Used to Clean and Monitor the Environment

Bacteria engineered to boost their hydrocarbon-degrading ability have been used for years to clean up crude oil spills at sea and on shore. In fact, these bacteria were the first living organisms to become the subject of a patent. Plant genetic engineering will undoubtedly also play a role in polluted soil reclamation. Many soil microorganisms, and even some plant species, have evolved to cope with pollutants of all kinds. Some bacteria use completely artificial organic compounds, which did not exist in nature as recently as sixty-years ago, as a carbon source for their own metabolic functions. These compounds are pollutants from chemical industries and dry cleaning practices, like trichlorethylene. These chlorine-containing organic chemicals are very toxic, except to the microbes that can degrade them. Several genes coding for chlorinated organic compound degradation have been cloned from these microbes and introduced into plants. The idea is not to engineer edible crop plants with these genes; the possible permanence of residues in these plants would make them unsuitable for human consumption. Rather, these plants would be used in polluted soils to take up the pollutants and degrade them, thereby cleaning these soils. A University of Washington–Seattle team has succeeded in producing fast-growing transgenic poplar trees that can degrade some chlorinated organic compounds. Lászlo Márton's team at the University of South Carolina–Columbia is engineering salty marsh plants for the same purpose.

Soils are polluted also by heavy metals like mercury, cadmium, and copper. Again, some bacterial and plant species have evolved genetic systems to detoxify heavy metals through a variety of means. Sometimes the resistance genes code for proteins that bind heavy metals and sequester them safely away, and sometimes they can reduce heavy metal salts to the metal itself—a much less toxic form of these pollutants. The experimental plant *Arabidopsis* and the yellow poplar have been engineered with some of these genes, and not only are they able to survive in the presence of toxic levels of heavy metals, they also detoxify these metals. These genetically modified plants extract the pollutants from the soil and concentrate them. These concentrates could then be incinerated under controlled conditions and eventually restore arable qualities to polluted soils. Plants engineered for pollutant reduction are at the experimental stage, and, like so many other research projects in biotechnology, have not yet been implemented.

Transgenic plants have also been used as biosensors (or bioindicators) of the environment. By definition, biosensors are cells or organisms whose function is to monitor environmental conditions. An early example of a crude biosensor was the canary in the coal mine. Coal miners used to carry a canary in a cage to detect the presence of carbon monoxide, a deadly poison. As soon as the canary collapsed, miners knew they had to evacuate that part of the mine to stay alive. Much more sophisticated biosensors, involving cells instead of whole animals, are currently being developed by the military to detect chemical and biological warfare agents present in trace amounts in the air. For obvious reasons, plants are good candidates to monitor soil contaminants. This approach has been used to determine the biological impact of radioactive pollutants present at low levels in the soils of Ukraine following the 1986 Chernobyl nuclear power plant disaster.

The explosion of Chernobyl's reactor 4 released huge amounts of radioactive cesium and strontium, two elements that are particularly pernicious because they are stably incorporated by living organisms in the place of potassium and calcium, respectively.

When present in cells, radioactive cesium and strontium are extremely mutagenic, as their beta and gamma radiations break DNA molecules. This breakage of DNA can induce cancers in humans. Therefore, it is extremely important to determine the biological effects of the radioactive cesium and strontium that was deposited in Ukrainian and neighboring soils after the accident. Direct measurements of cesium and strontium in polluted areas cannot give a precise indication of their mutagenic effects, especially in areas of low-level contamination, because of the complexity of the interactions between cesium, strontium, and soil components. How then can the biological impact of radioactive cesium and strontium be estimated in contaminated soils? This is where genetically engineered *Arabidopsis* plants proved essential.

A Swiss-Ukrainian team engineered *Arabidopsis* with a gene that is activated only after DNA breakage caused by ionizing (such as beta and gamma) radiation emitted by cesium and strontium. When activated, and after appropriate treatment of the test plants with dyes, the gene in the engineered plants caused them to develop blue sectors in areas of the leaves affected by the radiation. The more blue sectors appeared in the tissues, the more damage had been caused by the radiation. This provided a very sensitive way to measure the genetic consequences of the radiation. These engineered plants demonstrated that Chernobyl soils declared safe by the Ukrainian government (and subsequently reinhabited) do in fact contain low but biologically significant, deleterious amounts of radiation. Clearly, these engineered plants could be used as biosensors in other places where low levels of radioactive contamination are suspected. The Hanford Nuclear Reservation in the state of Washington, where vast amounts of radioactive waste are concentrated, should be a target.

A Most Unusual Application: Plastics-making Plants

The word *plastics* immediately suggests the image of giant factories where thousands of gallons of toxic solvents are churned together with noxious petroleum derivatives in huge metal vats at high temperatures. This image is true enough. Furthermore, making plastics involves not only petroleum-derived compounds, it also consumes vast amounts of energy that originate from the burning of fossil fuels. As we know, fossil fuels are bound to disappear completely one day, making the synthesis of plastics as currently performed impossible. In addition, petroleum-based plastics are presently an important source of pollution, because they do not degrade easily. On the other hand, we have become so dependent on plastics that a world without them is nearly inconceivable. What to do? Can this conundrum be solved? Let us think about what plastics are. At the chemical level, they are long molecules composed of a small number of subunits (often just one) that repeat thousands of times. The result is a fiberlike macromolecule, called a *polymer*. DNA, RNA, proteins, natural rubber, starch, and cellulose are all polymers, but their physical properties do not categorize them as plastics. Nonetheless, their existence proves that living cells are able to perform plasticlike polymerization reactions. Could cells, and in particular, plant cells, possibly make polymers with plastic properties? In other words, is it possible to engineer plant cells and coax them to make plastic nylonlike compounds? The answer is yes. Plants can make plastics after genetic transformation with specific bacterial genes. The bacterium *Ralstonia eutropha* carries genes that direct the polymerization of sugar into a class of plastics called polyhydroxyalkanoate (PHA). This is a perfectly natural phenomenon. *R. eutropha* grown in fermentation vats with sugar has in fact been used on a limited scale to produce PHA, which is utilized to manufacture shampoo bottles and disposable razors. One of the great advantages of PHA is that it is biodegradable and environment-friendly. However, the cost of producing PHA by bacterial fermentation is too high to make it competitive with petroleum-derived non-biodegradable plastics, at least for now. If PHA were to replace synthetic plastics in the very near future, a more economical method for mass production would have to be found. For example, since the sugar fermented by the bacteria that make PHA is extracted from plants such as corn, it is natural to imagine producing PHA directly in plants. This method would bypass the sugar extraction and fermentation steps, possibly reducing production costs.

The genes responsible for the polymerization of sugar into PHA were cloned from the bacterium *R. eutropha*, engineered for expression in plants, and introduced by transformation into the model plant *Arabidopsis*. These transgenic plants started making PHA granules by converting the natural plant compound acetyl coenzyme A. However, this experiment only demonstrated that the concept of plastic-synthesizing plants was not a dream: The small size of *Arabidopsis* plants precludes their use industrially. In 1994, Monsanto researchers introduced these bacterial genes into corn and started exploring the feasibility of growing PHA in plants on the larger, industrial scale. Their findings were disappointing: Even though the plants produced reasonable amounts of plastic, the energy cost of extracting the PHA from genetically modified corn was four times greater than synthesizing petroleum-based plastics. Since the greatest cost comes from the energy derived from the burning of fossil fuels, the manufacture of plastic in plants will not conserve these precious resources. Monsanto abandoned this project in 1999, but studies regarding energy consumption to produce *green plastics* continue elsewhere. This is not likely to be the end of plant-based plastics.

Is it possible to generate plastics in plants in such a way that further extraction and processing is not needed? This idea would work if plastic fibers were produced in tandem with natural fibers in the plant, as in the case of cotton, in order to produce a blended fiber directly in the plant. This objective has indeed been achieved with some preliminary success. Bacterial genes coding for the synthesis of the biodegradable plastic polyhydroxybutyrate (PHB, itself a subclass of PHA) have been introduced into cotton plants to produce PHB in the hollow core of cotton fibers as they grow on the plants. This could lead to the production of a "permanent press" type of cotton, in which core PHB fibers would prevent the wrinkling of surrounding cotton fibers. First results are encouraging, but the process must be fine-tuned, with the right proportion of PHB produced in the right place in the plants. Shirts and other articles of clothing made with this material are not yet on the market, and it is unclear whether they will be any time soon. Anyway, this idea is very intriguing: genes for the improvement of jeans!

We have seen in the last two chapters that biotechnology promises much, and realistically so, but has delivered little. Of all the possibilities described, only a few have been fully implemented, and these are mostly designed to protect crops against herbicides or insects. This makes sense from a business viewpoint; the companies that manufacture herbicides have a vested interest in developing crop plants that are resistant to the herbicides they also produce. This package guarantees that farmers who buy a given herbicide will also buy the seeds that are resistant to that herbicide. This allows for a kind of handshake profit structure between herbicide producers and seed companies, and it comes as no surprise that herbicide producers have acquired many seed companies. Basically, many farmers now purchase their seeds and their herbicides from the same corporation. This constitutes a purchasing simplification, but is this the best solution for the ultimate consumers, you and me? Perhaps not, because suddenly, competition is gone, replaced by pernicious monopoly.

On the other hand, less controversial applications seem stalled. Certainly, developing vitamin A-containing rice for consumption in developing countries is a worthy goal. Yet, we have seen that this project has been put on hold. And what about edible vaccines, aluminum detoxification, and phytoremediation? What about freezing tolerance induced by genuine plant genes? Again, it is hard to see what negative impact these applications could have. And again, these are applications developed by academic scientists, not by corporations. Is this why these applications have not yet been implemented? What will it take to allow plant biotechnology to be used in a true, benevolent fashion?[1]

László Márton's bioremediating plants, his detoxifying reeds, are still in the experimental stages. He has identified several new genes that decrease the toxicity of heavy metals, but the polluted salt marshes of South Carolina are not yet teeming with plants so engineered. It seems to me that restoring the health of the salt marshes is a worthy project, but it remains to be seen whether genetically modified plants will play any role in this endeavor. We will see next what types of societal issues and controversies surround the use of GM plants and how scientists and biotech companies have reacted to them.

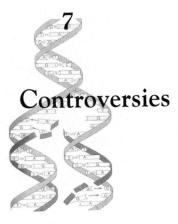

7
Controversies

IN THE EARLY 1980S, just a few years before the creation of the first herbicide- and insect-resistant crop plants, I was interviewed by a local newspaper reporter. He had heard about the new genetic technology, and somehow he had discovered my name and involvement in that field. He wanted to know what was in store for his readers, who were mostly central and eastern Washington state farmers. Trying to keep matters simple, I gave him two potential applications that I thought anyone could understand. First, I told him that one day, it would be possible to engineer strawberries for frost resistance. Second, I explained to him that, before too long, we would understand why weeds are generally much hardier than cultivated plants. Then, genetic engineers would clone the genes responsible for hardiness and transfer them to crop plants such as wheat, barley, and others. In this way, these plants would become more resistant to drought, viruses, and other environmental injuries. To my surprise, as I gave him my story, I noticed from the expression on his face that he thought I must have been totally insane to believe that anything like that would ever be possible. And indeed, his article, which he graciously sent me, announced that this crazy professor he had talked to was planning (in fact, I wasn't, and I had told him that) to grow frozen strawberries and turn barley into some kind of hardy weed. His readers must have had a good laugh. Well, people are no longer laughing, because plant genetic engineering is now a full reality. Furthermore, plant biotechnology faces possibly more opposition today, at least in some segments of society, than nuclear power plants.

We have heard much from opponents to genetic technology but we have heard hardly anything, at least in public, from academic scientists.[1] The latter have signed numerous petitions defending biotechnology, but they have communicated rarely with the public. We have heard even less from biotech companies except very recently, mostly through web sites and television ads. Why is it that some people are so passionately opposed to this technology? Are the negative claims justified? How have biotechnologists responded to public criticism? What role has industry played in the public perception of plant genetic engineering? And finally, what is the "second green revolution" promised by plant biotechnologists? Accusations of misinformation have been hurled repeatedly from both sides of the fence. The goal of this chapter is to address these questions and determine whether one can make sense of the struggle between the proponents and the opponents of the new biotechnology.

Opposition to Genetic Technology

In a famous article published in the British newspaper the *Daily Telegraph* in June 1998, Prince Charles of Wales accused plant biotechnologists of robbing God of one of His prerogatives, the genetic modification of crops. The immediate response of a panel of international scientists, published elsewhere, was to accuse the Prince of Wales of naïveté and to charge him with refusing food to the hungry. Their allusion to guillotined Queen Marie-Antoinette ("let them eat cake") of France must not have pleased the heir apparent to the throne of the United Kingdom.

My own view is that both declarations are equally absurd. Charles Windsor sees himself as an organic farmer (does he or does he not use Bt toxin; does he shovel manure over his fields?), and hence it appears that he believes he can make informed remarks about plant biotechnology.[2] In their rebuttal, the plant scientists made it sound as if the world is on the verge of collapse because of lack of food, yet they did not substantiate that point. Not a shred of scientific evidence was offered in either one of the articles. In fact, I believe that the absence of scientific information in declarations in favor or against biotechnology to be the main problem blurring the perception of plant biotechnology. The public has been uninformed and has, at best, heard a few brief analyses about the topic on television. People have also been exposed through the media to the ugly "Frankencorn" statue (a huge corn cob equipped with cobra-like fangs), built by Greenpeace adherents, and may or may not know that vandals regularly destroy test plots and greenhouses that most of the time do not even contain a single transgenic plant. People may also remember that riots in Seattle in December 1999 and demonstrations in Washington, D.C., in April 2000, both triggered by World Trade Organization conferences, had an antibiotechnology component. Both events resulted in destruction of property, injuries, and arrests. It is ironic, however, that neither conference discussed the applications of biotechnology and, instead, focused on issues of global trade. Apparently, some rioters and demonstrators believed that globalization of markets and plant biotechnology are one and the same thing.

Further, in June 2000, Jeremy Rifkin, the virulent opponent of genetic engineering and well-known (and politically savvy although not always objectively informed) American author of the famous 1977 book *Who Should Play God*, was invited by the Swiss government to participate in a public forum entitled "The Risks of Gene Technology: Phantom or Reality?" The forum degenerated into a shouting match when one of the scientists present, an unnamed Swiss Nobel laureate of Physiology, accused Rifkin of "playing on public fears" and talking "garbage." Rifkin, in turn, accused scientists of losing their objectivity given their ties to biotech companies.

What is the meaning of all this? How can the public sift through these conflicting pronouncements? Media sound bites have not helped. Let me offer an alternative approach based on science. First, opposition to plant biotechnology based on its violation of some divine order is illogical. We have seen earlier in the book that several crop plants were selected and propagated by humans, such as sterile, seedless watermelon, grapes, oranges, and bananas. These fruits were first produced decades ago. Plants bearing these fruits are evolutionary dead ends that do not survive without human intervention. Of course, one can always argue that it was the divinity's purpose to bless humans with these fruits and the ability to cultivate them. Following the same reasoning, one can also claim that it was the divinity's purpose to allow humans to invent genetic engineering. There is, of course, no end to this debate. Let's leave God out of the argument.

There are genuine concerns regarding what we put into our stomachs for survival. This is reasonable. Biotechnologists claim that genetic engineering is not new: Humans have done it for millennia, and cultivated wheat, corn, and tomatoes are there to prove it. There is an immense fallacy to this argument, however. True, plant breeding has been going on for millennia. But splicing bacterial genes into corn and canola is not exactly what Neolithic civilizations (that developed wheat, corn, and so forth) could have done. This type of argument, I think, has vitiated all discussion from the beginning. Some aspects of plant biotechnology *are* different and have nothing to do with conventional plant breeding. Had this been admitted earlier by biotechnologists, we would not be in this quagmire. On the other hand, plants can be engineered with plant genes, not just bacterial, fungal, fish, or viral genes. I believe few in the public would object to lettuce engineered with a nonallergenic celery gene, whereas inserting bacterial genes into plants may be seen as unnatural.

Oddly, genetic engineers have not told the public that in all likelihood, edible plants would be engineered mostly with plant genes or even synthetic genes in the future. One must realize that bacterial genes now used in agriculture were identified a decade or more ago, whereas similar genes in plants still await discovery. Plant biotechnologists' enthusiasm for the idea of introducing bacterial genes into plants, followed by a defensive attitude, has stacked the decks against public acceptance. Biotech companies, however, see huge potential profits, and their attitude must be differentiated from that of discovery-driven academic scientists. The arguments against the mixing of bacterial or other nonplant genes with edible plant genomes may be obsolete once plant genes are better understood. Even in the case of golden rice, we must remember that one of the two genes introduced into rice was actually a plant gene, and only one of them was of bacterial origin. I suspect that even *that* bacterial gene will be unnecessary once a plant equivalent is discovered and cloned. This is not to say that bacterial genes will be completely obsolete in future plant genetic engineering. But even so, stringent controls must be applied to ensure that the bacterial gene products are harmless. This, however, can only be accomplished by extensively analyzing the engineered plants for absence of toxic products, and that is exactly what was done in the case of transgenic soybean, canola, and corn: No toxic residues were found.

One example of the subtle use of a synthetic gene applies in the case of barley grain modified for chicken feed. Barley, unlike corn, is not a good source of food for chickens because the nutrients found in barley seeds are not readily digested by chickens. Barley has been engineered with a heavily modified (partially synthesized) bacterial gene in order to enable its seeds to produce nutrients that can be assimilated by the chickens' digestive system. This work was done by Diter von Wettstein's group at

Washington State University. Results show that this type of transgenic grain can save millions of dollars in areas that are corn grain importers, like Washington state, where barley, but not corn, is grown extensively. There is nothing wrong with biotechnology on its own, but we must be careful about what genes we introduce into crops and for what purpose these crops are going to be used.

Another public health concern deals with the antibiotic-resistance genes that biotechnologists introduce into genetically manipulated plants. Why is this done? It must be understood that any kind of transformation event, like the introduction of new genes into plants, is rather rare. In other words, when scientists use *Agrobacterium* or biolistics to introduce foreign genes into plants, most cells are untouched by the new gene. This shortfall is due to the inefficiency of our present techniques. The challenge is to recognize, or select, the few transformed plant cells from the massive numbers of other cells that have *not* incorporated the new gene. This can be done by attaching, or ligating, the specific gene of interest to an antibiotic-resistance gene. Since the two genes are physically combined, recognizing one will automatically allow recognition of the other. Plant cells transformed with the combination of the desired gene and the antibiotic-resistance gene can be selected by growing the DNA-treated plant cells in the presence of a lethal concentration of the antibiotic. The surviving cells will harbor the antibiotic-resistance gene *plus* the gene of interest linked to it. All other cells not containing the antibiotic-resistance gene (and not containing the gene of interest either) are killed by the antibiotic and will not clutter the culture plates from which engineered plants will be regenerated.

The presence of an antibiotic-resistance gene in genetically modified plants has been a valid concern. For example, the FlavrSavr® tomato contained a bacterial gene coding for resistance to the antibiotic kanamycin. Therefore, this tomato produced an enzyme that was able to destroy kanamycin. This antibiotic is not frequently used to rid humans of bacterial infections in the United States. However, it is still a useful antibiotic in human medicine—in Europe I myself have been treated with it for a kidney ailment. Kanamycin is normally administered by intramuscular injection over the course of several days. It certainly worked very well in my case. The objection to using a kanamycin-resistance gene in plants was that if one were treated with kanamycin for an infection, and if one had eaten FlavrSavr® in a salad for lunch, the enzyme responsible for kanamycin inactivation would be present in one's body. This activity might then interfere with the kanamycin if it had been given to fight an infection. In spite of extensive animal tests demonstrating that the kanamycin inactivation enzyme from FlavrSavr® was rapidly destroyed in the digestive tract, public concern was not alleviated. The outcome of this story is that antibiotic-resistance genes cannot be used to engineer plants without facing public opposition. This means that better transformation systems must be developed, in which gene transfer frequencies are so high that a selective agent, such as an antibiotic, is no longer necessary.

For now, some new techniques allow the breeding out of antibiotic-resistance genes, which then stop being a health concern. This task is done by breeding engineered plants containing an antibiotic-resistance gene with nonengineered plants and selecting in the offspring the plants that have kept the specific gene of interest but have lost the antibiotic-resistance gene. In addition, Swiss and Danish laboratories have demonstrated that selection schemes involving resistance to antibiotics may not be necessary after all: They have shown that the sugar mannose (not an antibiotic) can screen effectively for transgenic sugar beet and cassava. It remains to be seen whether this technique is applicable to other plant species. This is how mannose-based selection of transgenic plants works. We have discussed how gene transfer to plant cells via *Agrobacterium* or biolistics must be followed by a step involving regeneration of cells into green plants. This step is carried out in culture medium containing sucrose, regular table sugar. Another type of sugar, mannose, interferes with this regeneration process, because plants cannot utilize mannose as a source of energy. Researchers reasoned that plants made transgenic with a gene called *pmi* that directs the metabolism of mannose into glucose (which, like sucrose, can be used as an energy source) would survive in the presence of mannose. This assertion proved to be correct: Plant cells transformed with the *E. coli pmi* gene can indeed utilize mannose for metabolism because they first convert mannose into glucose. This *pmi* gene has nothing whatsoever to do with antibiotics, alleviating the concerns discussed earlier. This gene is also naturally present in the thousands of billions of *E. coli* cells that colonize the human intestine. There *are* ways to avert the use of antibiotic-resistance genes. It should be noted that soybeans resistant to Roundup® and canola resistant to Liberty® do not contain such antibiotic-resistance genes. In this case, the herbicide itself was the selective agent in the transformation and selection process. It helps to know these differences.

What about the case of plant food quality? As we have described, engineering rice for provitamin A content is a worthy and inoffensive goal (one cannot overdose on provitamin A). Nevertheless, genetic engineers would benefit from discussions with people involved in Third World nations' development. It turns out that provitamin A in regular rice is lost if the rice grains are polished. However, many poor people in the Third World do not polish their rice because of the cost or time involved and, hence, benefit from provitamin A present in the seed coat of unpolished rice. On the other hand, many poor or landless Third World people cannot afford to grow rice. They grow and eat barley instead, and many others grow nothing at all. What this analysis means is that golden rice should be made available to carefully selected populations, without violating cultural practices. In other words, we should not concentrate all of Asia into a single category: the not-so-poor polished rice eaters who would definitely benefit from golden rice. Yet, one should also keep in mind those who cannot afford to grow rice and those who, for generations, have cultivated an inferior source of grain: barley. Meritorious as it is, the golden rice effort still has to be put into a cultural and sociological context.

However, boosting vitamins or lowering fatty acid content in plants is a significantly different matter, I think. Academic and corporate scientists alike are motivated by monetary grants (in academia) and profits (in industry) to do highly visible research. Grants and profits ensure salary raises, job stability, and promotion, which in academia largely depend on decisions made by financially motivated administrators. An example of this is when nutritional fads interfere with research goals. Fads can drive perceived needs, needs drive motivation, and motivation drives salaries upward. Vitamin- and other micronutrients-containing pills together with a diet low in saturated fatty acids have been touted as salubrious elements in a long, healthy life. Vitamin and other supplements are advertised everywhere, whereas everyone knows that lard (which contains high concentrations of saturated fatty acids) is anathema to good health. Do these fads represent reality? Excess vitamin A and D as well as excess calcium and iron can be toxic, but a reasonable dose of lard is unlikely to be so. After all, many of the very old people in the Western world had no access most of their lives to vitamin supplements. In fact, part of their daily diet included bacon (from which lard is derived) and even butter and eggs (cholesterol). Granted, vitamins are probably better than lard, but is this a life and death issue? I think not. The question is: Do we really need vitamin-enriched lentils or broccoli and vegetable cooking oil low in saturated fats? Is it really necessary for biotech companies to make all vegetables vitamin-enriched?

This brings up the question of overdosing. What about the effects of high vitamin or mineral doses on growing children? There is the (nonvitamin) example of children made seriously sick after their parents put them on a faddish low-cholesterol diet—one potentially lethal for infants. I believe good common sense is necessary, and yet, some scientists may have lacked it when asked to advise the public. They may have behaved in such a way because of overenthusiasm or for pecuniary reasons. Public skepticism is definitely warranted. As for the minerals iron, calcium, selenium, and iodine, the same caution applies: These elements are toxic when taken in large doses, and enriching plant material with these elements could be very tricky. Popeye the Sailor may have needed an extra dose of spinach to defeat his enemies, but his spinach was not spiked with supplemental iron. In addition, why would anyone want to eat a genetically engineered, selenium-enriched avocado when vitamins and micronutrients can be ingested in pills obtained over the counter in any grocery store?

Another objection to genetically modified crops deals with the potential transfer of herbicide-resistance genes to sexually compatible, wild relatives of cultivated plants. This seems to have happened, to a limited extent, in experimental plots in England. However, after several years of use of Roundup Ready® soybean in the United States and the same type of practice in Canada with Liberty®-resistant canola, there have been no reports of the appearance of herbicide-resistant "superweeds" invading fields. Not much of a threat exists yet. The same thing goes for corn and cotton engineered for insect resistance with Bt toxin. However, there is more at stake than gene transmission to weeds. When insects feed on anything that is deleterious to them, rare individuals will invariably mutate and develop tolerance to it. For example, the once-powerful insecticide DDT lost its effectiveness a long time ago. So far, and after just a few years of experience with insect-resistant crops, nothing similar has occurred. Undoubtedly, however, Bt-resistant insects will appear, in time, proliferate, and render the current transgenic crops inefficient. Still, there are many types of Bt toxins, and it is very possible that once resistance to one kind of Bt toxin develops, plants could be reengineered with another Bt toxin variant. Nevertheless, the number of different Bt toxin genes present in nature is not infinite, so sooner or later other strategies will be necessary. We do know that many farmers, including some organic farmers, have been spraying the Bt toxin protein on their crops for several decades, without much public opposition and apparently without ill effect. Today, a precautionary measure against the spread of resistant insects is now routinely taken by farmers. This measure consists in planting alternating rows of Bt-producing plants with regular, nonengineered plants called *refugia*. The principle at work is that insects do not know the difference between transgenic and nontransgenic plants and will invade both types indiscriminately, thereby decreasing the evolutionary selection pressure that would favor resistant insect genotypes. Returning to the question of transgene transfer to sexually compatible plant species, it is unsettling to note that intact Bt toxin genes, clearly engineered with chimeric promoter and terminator, were found in native corn varieties in southern Mexico. These samples were collected in the year 2000 in the remote mountains of Oaxaca province where no engineered corn was supposed to have been cultivated. This indicates that, somehow, pollen from engineered corn grown elsewhere in Mexico was able to travel very significant distances and fertilize native varieties. Even worse, Mexico had implemented a moratorium on the cultivation of Bt corn in 1998. It is not known whether these native corn varieties accidentally carrying the Bt toxin gene have had any impact on the local insect populations.

Other issues surround the use of corn and cotton engineered with the Bt toxin gene, as one can well imagine. A recent report from Cornell University that Monarch butterfly caterpillars have been killed by feeding on insect-resistant corn pollen has stirred great anxiety. The Monarch butterfly is indeed a somewhat at-risk species, and as it is not a pest for corn, it was never meant to be

hurt by these transgenic plants. The Cornell report has been criticized for not representing true field conditions, and there is as
yet no evidence that Monarch butterflies are being wiped out by
transgenic corn. Subsequent field and laboratory studies conducted at the University of Illinois found no relationship between
the mortality of common butterfly caterpillars (black swallowtail
caterpillars) fed with corn pollen engineered with a Bt toxin gene.
Finally, in summer 2001, the Environmental Protection Agency
(EPA) concluded that the impact of Bt corn on Monarch butterfly caterpillars is very small: Only 1 in 100,000 insects are affected, and even the affected larvae can develop into healthy adults.
The EPA has not ruled out that another endangered butterfly
species, the karner blue butterfly of Wisconsin, could be affected
by the consumption of Bt corn pollen.

Nevertheless, the possible effects on unintended targets, the
Monarch butterfly and the Wisconsin butterfly, raise the question
of such effects on other unintended targets: humans. Many people are concerned about the insertion of allergens into plants that
are not known otherwise to contain them. This is a legitimate
concern. Extensive studies show that transgenic corn and transgenic canola do not contain new allergens due to their genetically
altered status. This does not mean that enhanced allergenicity in
other transgenic plants cannot occur. Public concern about new
allergens (and other health issues) would be alleviated if a simple
measure were taken: the labeling of transgenic plants and products derived from them. People would then be able to choose
whether or not to consume food items containing foreign genes.
There is a big stumbling block, however.

Labeling has been staunchly opposed for years by biotech
companies in the United States. This obstruction is very strange
in a country where even the nutritional value of bottled distilled
water is described at length on labels (it is 0 percent for all categories). The absence of labels is perfectly legal so far, except
when known allergens have been added to the product; but pub-

lic skepticism about the absence of labeling has led to a belief in some "conspiracy theory" surrounding GMOs. Indeed, one might ask, "If they have nothing to hide, then why are they hiding it?" The biotech companies' response has been, "What should we label *for*?" This type of response fuels suspicion, because it suggests that companies do not know what the effects of a transgene may be. This is not the case. Labels could contain information like "engineered with a Bt gene" or "engineered with a glufosinate-resistance gene." An educated public should be able to understand that. This area is where biotech companies have failed miserably. Many scientists think that the arrogance of companies sparked the debate right from the start. No efforts were made to tell people what plant genetic engineering is and what transgenes do. Companies are now involved in rear-guard battles, because they never went to the trouble of educating the public in the first place. A television commercial that started airing in May 2000 is attempting to redress the situation, but it is remarkably uninformative, unfortunately. It conveys a "feel-good" message that is about as convincing as the advocacy of the purchase of a cemetery plot. I think the public deserves better. The onus is on biotech companies. After all, it's *their* business.

In addition, there are health concerns about possible hidden effects of transgenes. The notion is that a foreign gene could insert itself into plant DNA in such a location, or have such an effect, that it would disrupt a normal metabolic pathway and turn it into a toxic pathway. This question is particularly difficult to address because it is not specific. In other words, a transgene can be accused of potentially having an unknown, unpredictable effect. How does one predict the unpredictable? One cannot. It turns out that indeed, transgene insertion into experimental non-food plants had an unpredicted effect in one set of experiments. Researchers discovered that a transgene supposed to direct the synthesis of a given fatty acid also led to the production of an unanticipated, *other* type of fatty acid with undesirable properties. However, transgenic foods are analyzed extensively for the presence of known toxic compounds, including allergens, and so far, none have been found in the engineered plants we eat. In most cases, the disruption of any metabolic pathway in all likelihood would be lethal or severely damaging for the plant. Genetic engineers take great care in identifying transgenic plants that are perfectly healthy and that retain all the qualities of the plant before it is engineered.

We have seen earlier that gene targeting in plants, that is to say, the insertion of a transgene in a predetermined spot of a plant genome, is still in its infancy. Once this technique is refined, it will be possible to insert foreign genes with great accuracy. We will know with the highest confidence that this process will have no deleterious effects on plant metabolic pathways. This certainty will also be made possible once the full sequences of crop plant genomes are known—in just a few years from now. And finally, there are many federal regulations that biotech companies must follow before they can release any genetically modified food product. The Food and Drug Administration (FDA), the Environment Protection Agency (EPA), and the U.S. Department of Agriculture (USDA) are responsible for monitoring these products.

Unfortunately, even multipronged monitoring cannot prevent human error. Biotech companies are not the only actors in the long chain of events that leads to the consumer. Farmers, distributors, and food product manufacturers are also very much involved. The potential for human error leading to, for example, foodstuff contamination with unauthorized genetically altered products increases with each step. This is exactly what happened in October 2000 in several Western states. Antibiotech groups ordered the analysis of taco and tostada shells made from corn, produced by Mission Foods and distributed by Western Family Foods in Oregon, Washington, California, Utah, Idaho, Hawaii, and Alaska. This study revealed the presence in these products of a Cry9C Bt toxin variant that had not been approved by the FDA for human consumption. Cry9C had been approved for use in animal feed and is grown by about 3,000 farmers, but it was not approved for human consumption. Western Family promptly recalled these tostada and taco shells, but it is not known if any were sold to customers. It is thought that the presence of unapproved genetically modified corn in taco shells was due to unintentional contamination of regular corn with StarLink, the Cry9C Bt toxin gene-containing variety produced by the biotech company Aventis CropScience. Most of this corn has now been traced by government agencies, but about 1.2 million bushels (1.5 percent of the total StarLink grown in 2000) remain unaccounted for. In a case like this, neither Aventis CropScience nor the FDA and the EPA can be blamed for this contamination. Further, since genetically modified corn is indistinguishable from regular corn when grown, harvested, and processed, it is understandable that a mix-up was waiting to take place. On the bright side, if there is one, StarLink corn is extremely unlikely to make anyone sick. Still, this incident is unacceptable and demonstrates that stringent regulations, coupled with frequent analysis of food products, are necessary to establish a strong demarcation between crop plants approved for human consumption and those that are not.

Last but not least, there is concern that massive application of genetic engineering would narrow cultivated plants' gene pools. At the present time, different varieties of edible plants are cultivated in different parts of the world. For example, the *japonica* rice variety is preferred in Japan while the *indica* variety is much more popular in South Asia. Similarly, potatoes grown in the United States are not identical to European potatoes. This divergence offers a great advantage because different plant varieties react differently to pathogens. There have been cases in history of entire crops wiped out by disease, causing great misery. The Irish potato famine is a well-known example. Another example is

that of the French wine industry that was devastated in the 1920s
by an epidemic of mildew. The Irish potato blight did not destroy
all potatoes in the world, and the mildew epidemic in France was
stopped by the grafting of mildew-resistant American grapevines
onto French grapevine stocks. Avoiding further spread of these
epidemics was possible because not all potato and grapevine varieties were equally susceptible to pathogens. This is what is meant
by the phrase *gene pool*: the existence of different cultivated
plant varieties with subtle genetic differences that give them natural resistance to various pests and diseases. However, if massive
engineering of some crops with a given transgene were to occur,
it would be unlikely that all possible varieties of those crops could
be engineered. This would be too time- and money-consuming.
In a scenario where only one crop variety was genetically engineered, this variety could supplant all others, making that one
dominant variety profoundly vulnerable. Such practice would
narrow the gene pool so that it could no longer prevent a catastrophic epidemic, as no naturally resistant varieties would be left
in existence. This is a very serious issue that biotech companies
must address.

Many concerns expressed by the public have a valid base. I
believe these concerns must be answered and that they are
answerable. My own general impression of public opinion in the
United States is that genetically modified food plants are not
viewed as evil, within certain limits. However, conversations on
this topic with my undergraduate students (who tend to be more
candid than graduate students) have shown overwhelmingly that
absence of labeling of genetically modified food products is seen
as an intolerable infringement upon their freedom of choice. I
must agree with this opinion. In early 2000, the Clinton administration suggested timidly that perhaps biotech companies should
start to label their products on a voluntary basis. It remains to be
seen whether this will happen. Note that Japan has recently
decided to make labeling of GM foods mandatory, something that

has triggered the ire of major U.S. biotech companies since their export market to Japan may be impaired. Nonetheless, a move towards making the labeling of bioengineered foods mandatory is shaping up in the United States. In late 2000, a bill called "The Genetically Engineered Food Right to Know Act" was introduced in both houses of the U.S. Congress. This bill would require the labeling of food that contains a genetically engineered material, or was produced with a genetically engineered material. Critics of this bill point out that it contradicts existing laws that regulate food safety. To use a single example, that of the Bt toxin present in GM corn, no law stipulates that *regular* corn sprayed with solutions of Bt toxin protein must be labeled, because the EPA considers the Bt toxin safe for human consumption. Then, declare the critics, why should bioengineered corn be labeled at all? Indeed, if applying Bt toxin protein externally is considered legally safe, plants making their own Bt toxin protein internally, through genetic engineering, should be considered safe and should not be labeled. This is a compellingly logical argument, but will it convince the public? Some even go as far as to say that in the absence of health and safety concerns (presumably as determined by the USDA, EPA, and FDA), requiring the labeling of GM foods might violate the First Amendment (the right to free speech) of producers of bioengineered foods. Clearly, many legal battles loom on the horizon, and this contentiousness—corporate and constitutional lawyers deciding what types of foods end up on our plates—is probably not what a concerned public wants.

At this point, it is worthwhile to discuss the reactions of the European public to genetically modified crops. Europeans are much more adamantly opposed to GMOs than Americans. I think there are many reasons for this attitude. First, Europeans do not trust their elected officials. This state of affairs is also partially true in the United States, but in Europe government instability is chronic and endemic. Second, multinational corporations (such as the ones that produced transgenic corn and canola) are seen as tentacular giants acting in concert with the government and, in collusion with the latter, doing everything they can to impose their products on a powerless public. Politicians and company CEOs are seen as equally corrupt. Why then trust what they say about the safety of transgenic crop plants? Furthermore, many European countries have fine-tuned cuisines that local citizens think need no improvement whatsoever. *Spaghetti alla matriciana* cannot possibly be made better by fungus-resistant wheat and virus-resistant tomatoes. Similarly, *cassoulet de Lyon* will not benefit from transgenic beans resistant to some herbicide, and *Sauerkraut mit weiss Wein* will not be made better by the use of transgenic, vitamin E-rich cabbage. Why tamper with something that is already perfect? And indeed, how does one defeat such an argument? Surely, someone will profit from cornering the market with transgenic plants but it will not be the food itself (which by definition is already magnificent) nor the public, because prices at the grocery store will go up no matter what. Last but not least, Europeans have never adopted the idea that new is necessarily better; they have long-standing traditions and they cling to them. Thus, Europeans can be seen as a skeptical lot as far as plant genetic engineering is concerned.

On the other hand, a scientific survey conducted in Europe and in the United States in 1998 shows that the European public is more aware of the basic nature of biotechnology than the American public. This greater awareness is accompanied by a much more negative perception. The authors of the survey do not invoke the culinary arguments I mentioned here to explain this observation. They stress that greater media coverage of biotechnology in Europe is associated with greater public concern, that Europeans distrust their regulatory authorities, and that they have better textbook knowledge of biology and genetics than Americans. Their conclusion is that Europeans see biotechnology as much more threatening than Americans. The authors also raise the following question: "How should science, industry, and governments respond?" And indeed, if more knowledge of biotechnology means less acceptance, does it mean that the education given the European public was flawed or biased? Does this mean that better information is necessary? These questions were not addressed. At any rate, let us not dismiss the European response too quickly; perhaps the Marquis de Lafayette's ghost is talking to us. Given the raging controversy on the European continent, the European Union has put a temporary ban on the import of GM foods from the United States.

Biotechnologists Respond

We have seen that scientists have understood the potential risks of biotechnology since 1975. Academic scientists must today obtain an approved Memorandum of Understanding and Agreement from their universities before they can engage in federally funded research programs that involve recombinant DNA and transgenic organisms. This precaution is in place to ensure that no federal guidelines are violated. Biotech companies must run a battery of safety and other tests, to be verified and approved by the FDA, before they release a commercial genetically engineered product. It would seem that the federal government determines what can and cannot be marketed. Yet some sector of the public is not convinced. Does this mean that the public is suspicious of science? The answer is yes. Since the U.S. public has little awareness of the science that underlies biotechnology, will that same public allow a more objective education of the consumer? This is perhaps what biotech companies are trying to achieve. On the other hand, these companies are viewed less than favorably by a portion of the public, and it is now up to academic scientists to initiate solid public debate. Whether scientists can be convinced to do that is another story.

In addition to concerns about the toxicity of genetically engineered corn, soybean, and canola, there are questions about the creation of superweeds. As we saw in previous pages, this has not happened after several years of growing herbicide-resistant plants. This fact does not mean it will never happen. To confront this threat, a new approach to genetic engineering for herbicide resistance has been proposed by Henri Daniell working at Central Florida University. This approach consists in engineering chloroplast DNA rather than nuclear DNA for herbicide resistance. Chloroplasts are the cellular bodies that perform photosynthesis. They contain their own short strands of DNA, separate and distinct from the vast majority of the DNA, which is located in the cell nucleus. When plants are pollinated, the pollen grains, or male germ cells, are dispersed by the wind and insects. Female germ cells, or ovules, are held in place securely inside the ovary, which is not mobile. Thus, pollen grains carrying transgenes can be dispersed; ovules cannot. It turns out that pollen grains do not contain chloroplasts. Daniell questioned, why not clone transgenes, like herbicide-resistance genes, inside chloroplast DNA? In this way, the transgenes would not be present within pollen grains, which contain nuclear DNA but no chloroplast DNA. Transgenes would not be dispersible by way of the pollen, and cross-pollination of weeds would not be possible. It worked. Plants that possess genetically engineered chloroplast DNA do not donate their transgene to other plants through their pollen. It remains to be seen to what extent this approach will be used in the field. It shows, however, that science can often find solutions to difficult questions. As for the question of biotechnology narrowing a gene pool, biotech companies have not taken action to resolve this yet.

The Next Green Revolution
Biotechnology enthusiasts see a second green revolution as a necessary step to save the planet. What was the first green revolution and what did it achieve? The green revolution consisted of massive plant breeding programs in the 1960s and 1970s, aimed at

enhancing the qualities of crop plants grown mostly in the Third World. The philosophy of the program was that the development of new, high-quality hybrid varieties would help eradicate hunger in those regions. The programs have been very successful in some respects; nations like India that used to be food importers now export vast amounts of grain. What is more, the new crop varieties were distributed nearly free to countries in need. This generosity was possible because international organizations, not companies, were in charge of the breeding programs and their supervision. Biotech companies and others are now advocating a repeat of the first green revolution, this time based on genetic engineering. This looks like a very worthy goal, but the conditions that surrounded the first green revolution have significantly changed this time around. We will see some implications of this new situation later in this chapter.

Monopolies

Academic science made biotechnology possible, but applications of biotechnology to practical uses like herbicide resistance in plants were all developed by industrial companies. This was to be expected since university laboratories are not equipped to generate marketable goods, which is not their function (except in the case of land-grant universities that may not be able to compete with private companies much longer). We live in a world where mergers and acquisitions of companies are commonplace. Megacorporations are constantly trying to monopolize markets, and plant biotechnology companies have not escaped this. Sometimes I wonder if this drive is partially responsible for the negative perception of biotechnology. This is a political issue, not a scientific one. However, could it be that the public feels manipulated at its most basic level: food? It may seem that "blind" corporations whose sole goal is to make profits are interfering with a basic human need. Not so long ago, farmers were the primary producers of our food. This situation is rapidly changing. Farmers themselves, unless they oppose biotechnology, now obtain seeds from companies that not only engineer seeds, they also produce them. Large biotech companies have merged and acquired seed companies that were once independent. Today, farmers buy their seeds and herbicide as a single package, from a single company. This packaging certainly simplifies farmers' lives. On the other hand, farmers really have no choice, and this restriction is not exactly in line with the U.S. traditions of free choice and business competition.

As for the consumer, how have we benefited from plant biotechnology? Setting aside issues of safety (which I think are presently nonexistent, as discussed earlier), what differences exist in engineered breakfast corn flakes (made from insect-resistant corn) and cooking oil (made from herbicide-resistant canola)? Do they taste better? Do they cost less? I doubt that your corn flakes and cooking oil taste better or are less expensive today than they were before biotechnology. Then, who benefits, who makes the profits? I believe biotech companies profit more than anyone else. But then, why should the public be pacified with remarks concerning the safety and advantages of genetically engineered crop plants when there are no discernible differences between engineered and traditionally farmed foods? Rather than just saying that engineered food plants are good, perhaps companies should lower their profit margin and reduce the prices of bioengineered products. That deed is unlikely. The issues remain complicated. Recall the upside: the use of herbicide- and insect-resistant crop plant varieties has led to a decrease in herbicide and pesticide use and a concomitant decrease in chemical loads in soils. Nevertheless, the embedded costs of the manufacture of the pest- and herbicide-resistant plants may keep the prices of end products high.

Academic scientists, at least those not tied by contract to biotech companies, have started to retaliate. One of the creators of the high-yield transgenic rice varieties, Maurice Ku, has publicly declared that he wants to make these plants available to everyone at no cost. To accomplish this feat, he and his colleagues at Japan's Nagoya University have patented their engineered varieties. This move, they claim, would prevent biotech companies from patenting these plants themselves. This ploy might work indeed, but some argue that university researchers are not so fashioned to control the distribution and other business aspects of transgenic plants. Nevertheless, it is interesting to note that some scientists are not hesitating to take the high ground and relinquish potentially enormous profits.

What About the Third World?

Returning to a new green revolution, what role will biotech companies have? As mentioned earlier, the first green revolution was sponsored by nonprofit organizations. The Rockefeller Foundation was prominent among them; yet, today it is reluctant to continue the funding for golden rice research. If indeed the Third World is in need of a second green revolution, from where will the funding come? But first, let us examine that need. Its advocates present a second green revolution as being driven by the population explosion. Yet, experts disagree as to what the magnitude of this explosion will be. Some, including the Indian Nobel laureate for Economics, Amartya Sen, claim that the food production-population ratio is not necessarily an indication of whether famines will occur. Many other factors are involved; famines have occurred in Asia in periods of high agricultural output and were due to unemployment, speculative stockpiling, and low purchasing power, or poverty, not food shortages. The poorest of the poor often starve or are undernourished because they cannot afford to buy food, not because the food is not *there*. One main problem is distribution and access. There is nothing biotechnology or even a green revolution can do about these factors; these are socioeconomic and political problems.

The first green revolution was not able to eradicate hunger. There is even evidence that hunger has worsened in some segments of the Third World population because of it. The first green revolution crop plants had much higher yields, but they also demanded heavy use of expensive agrochemicals. Not all farmers were able to afford these. As a result, after being driven out by richer farmers, poorer farmers migrated to urban slums where they grew even poorer than before. We cannot consider green revolutions lightly; they are not a cure-all for the world's ills. Finally, not everybody agrees that total plant food production should be increased beyond levels that can be achieved by conventional breeding, because some claim agricultural output (without biotechnology) will be able to meet demand for at least the foreseeable future.

Even in the event that a second green revolution is needed, by whom will it be implemented? Developing countries such as India and China have very active plant biotechnology programs. However, most of the patents are held by Western corporations. This makes it impossible for developing countries like India and China to develop transgenic food plants and commercialize them without breaking international laws.[3] What then can be done? Ismail Serageldin, a World Bank official, does not see the private sector investing in the needs of developing countries; monetary returns would be too low. He recognizes that intellectual property rights and patents must be respected, but he realizes also that developing countries should not be neglected if they apply genetic technology to their own food production. As a potential solution, he proposes that Western biotech companies form partnerships with the public sector in developing countries. Profits for companies would come from royalties, not through the monopolization of markets. Monsanto has already established such consortia with Mexican and Kenyan labs.

Further, the Rockefeller Foundation, a nonprofit organization, is a major supporter of the Canberra, Australia-based CAMBIA (Center for the Application of Molecular Biology to International Agriculture). There, the emphasis is on crop plants grown in developing countries, but the focus is not on adding single genes to plants, as is currently done by individual companies. Instead, this institute has taken what it calls a holistic approach to determining gene activity in plants and tailored them according to local needs. For example, CAMBIA has helped Chinese researchers develop a transgenic rice variety with a prolonged grain-filling period, the result of which is a yield increase of 40 percent. It is still unclear how CAMBIA will deal with intellectual property rights, but its core principles seem to work excellently. Meanwhile, crops such as cassava (a plant with a large, starchy root used as a staple food in the tropics), bananas, the sweet potato, and oil palm trees are receiving attention thanks to the establishment of consortia between developing countries' institutions and Western universities. This cooperation would not have occurred if biotech companies had been involved in only these applications, because the crops just described are considered to be small crops, not worth much investment in research and development. But here again, patent issues remain. These conflicts might be solved, given time, as developing countries become more affluent and start to generate their own patents.

China, with its population of 1.35 billion, has released a variety of transgenic crop plants apparently engineered with the Bt toxin gene. According to my contacts in China, this was done under a license provided by Western companies. As for the public perception in China of GM foods, it seems to be positive. This is not surprising in a country that must feed its huge population. The Chinese government has maintained a very positive attitude toward high technology and has promoted the idea of genetically engineered food plants for several years. On the other hand, given the tight government control over the media, there never has been much debate on the potential hazards of GM foods. And indeed, other contacts I have in China are not aware that they are consuming transgenic plants. In a cynical sense, China has become the largest experiment testing the effects of transgenic food plants on human health.

Conclusions

The projected threats of plant biotechnology against humanity have not come to pass. There is no scientific evidence that engineered corn, soybean, or canola have had a detrimental impact on humans and the environment. Americans and Canadians are not suffering short-term or medium-term effects from the consumption of these transgenic foods. Nevertheless, biotechnology is not exactly popular in North America and elsewhere. We have seen that there are many reasons for this attitude. These range from the notion that humans should not tamper with nature, to fear for public health, to outrage at the lack of labeling for transgenic foodstuffs, to Western neoimperialism, to biotech companies being monopolies, and even perhaps to their creating needs that do not exist. Most of these arguments are emotional, but that does not mean that they are not justified. Biotech companies must now face the challenge and demonstrate that they are no longer purely profit-motivated, but that they also want to make the world better.

Some aspects of plant biotechnology are quite positive. Edible vaccines, soil reclamation through phytoremediation, green plastics, and golden rice (if ever produced and distributed wisely) seem not only inoffensive, but truly beneficial. Other applications, like plants engineered to follow faddish diets, may not be so innocuous, in which case consumers ought to be given the choice to purchase these vegetables or not and need to be well informed. This means visible labeling in no uncertain terms. Further, biotech companies and the FDA should make the results of their food safety tests fully available to the public, including raw data. Short of that, I do not believe the consumer will ever be convinced that genetically modified foods are safe to eat. Such a move would also demonstrate honesty and openness on the part of the government and private industries.

Meanwhile, extremist groups have destroyed research facilities and "Frankencorn" (that epitomizes all "Frankenfoods") is paraded on the streets of our cities to incite fear. The Greenpeace movement has distributed pamphlets claiming that GM food plants are untested, which is untrue, and that consumers are used as guinea pigs. A powerful rebuttal of this position is provided in the 1996 book *Herbicide-resistant Crops* by biologist Stephen O. Duke that demonstrates how herbicide-resistant crops, among others, are subjected to stringent regulatory mechanisms, risk assessment, and environmental studies before they can be released. The FDA's and other web sites show the same thing over and over again. Nevertheless, publishing books, building home pages, and yelling in the streets does not constitute a debate. When will a sane, public deliberation take place?

Biotech companies have so far reacted to public opposition and concerns sluggishly, assured that the law is on their side. Hypocrisy and duplicity, however, are apparently not beneath them. For example, Gerber has decided to ban all transgenic products from its baby foods. This sounds like a courageous (and probably financially rewarding) move until one learns that Gerber is owned by Novartis, a major plant biotechnology company. To my knowledge, Novartis has not yet released any transgenic crop varieties, but their move could be construed as a purposeful profit strategy to defeat the competition by barring the use of one of its own potential products in another one of its own products, thus perhaps insinuating that competitors' products may contain GMOs.

Even worse, recently, in March 2001, opposition to GM wheat in North Dakota triggered a downright nasty reaction on the part of Monsanto, the creator of a new Roundup®-resistant wheat variety. North Dakota produces and exports vast amounts of wheat. It is in the interest of Monsanto to persuade North Dakotan farmers to use their engineered wheat on as large a scale as possible. However, the farmers were well aware that their international customers (including Canada) might not purchase that wheat or even their nonengineered wheat, for fear of contamination. The North Dakota legislature has asked Monsanto for a two-year moratorium on the introduction of Roundup-Ready® wheat. Monsanto's reaction was quick and simple: The company threatened to stop its funding of the ongoing research on all GM wheat performed in the state. This would severely hamper agricultural research in the whole state and adversely affect its fiscal balance sheet. This shows again that corporate profits often directly clash with public concerns and potential scientific developments. Instead of threats, Monsanto could have used this opportunity to provide public education.

Unfortunately, nobody seems to care enough about the science of plant transgenesis to explain it to the public. This must change, because the bottom line is with the consumer: If he or she decides not to buy transgenic products, they will not be sold. This is what is going to happen if the public remains uneducated about these issues and continues to be manipulated by multinational corporations.

It might be argued that my ethical problems with biotech companies are as naïve and as subjective as Prince Charles's theistic opposition to biotechnology. After all, industrial research and development funds have made possible the synthesis and commercialization of products such as antibiotics and plastics. These are regarded as good by most people because they make life easier and better for just about everybody on the planet. However, I find it impossible to characterize herbicide-tolerant and insect-resistant crop plants in the same way. These products are not making the life of most of us easier and better. Corn is not more abundant, cheaper, or better than before biotechnology, and neither is canola oil nor tofu. So far, applied plant biotechnology has been neutral (or negative if one considers the backlash against it) in what it has offered to the public. It is legitimate to ask what the vision of biotech corporations really *is*.

Finally, on an optimistic note, I cannot stress enough the impact that genetic manipulation has had and still has on basic plant biology. This also holds true for the field of human medicine. Recombinant DNA and gene transfer techniques have totally changed the way we study living systems and understand them. As *Time* magazine put it once, gene technology is indeed an "awesome skill." It is up to us to use it wisely.

Epilogue

The sixth century B.C. Greek fabulist Esopus described the human tongue as the best and the worst of things. I think biotechnology can be seen that way, too. Engineering plants with plant genes is as close as we can get to traditional plant breeding and should elicit little negative response from the public. We have seen that overexpressing resident plant genes can prevent accumulation in roots of toxic salts. More recently, scientists have demonstrated that introducing an *Arabidopsis* gene determining dwarfness into basmati rice significantly shortens the stalks of the plants, allowing more energy input into the grain and more starch deposition. This improvement would not have been possible through classical breeding. Soil reclamation by phytoremediation is also a worthwhile goal. These, I think, are useful applications of genetic engineering. I believe that other applications like frost and disease resistance will follow once plant genetic pathways are better understood and new genes are discovered.

Biotechnology could have a bright future in the field of ornamental plants. Wouldn't it be nice to have roses or gladioli that glow in the dark? This could be achieved by engineering flowering plants with the firefly *luc* gene. In fact, carrot and tobacco plants engineered with the *luc* gene were produced as early as 1986. Their glow was not visible to the human eye but could be recorded with sensitive photographic film. Further engineering might make the glow visible. And what about flowers that could change color with temperature? My colleague, Edgardo Filippone of the University of Naples, Italy, once suggested this to me. He proposed that cut flowers could be engineered to respond to body heat. This, he said, would prove one's love for the significant other: "Darling, these daffodils I'm offering you are bright red because I held them close to my heart, which is burning with love for you!" Once the ardor of the moment had passed, the daffodils, now in a vase, would return to their normal yellow and white color because of the temperature drop. Such applications of biotechnology for horticultural purposes are not as silly as they may seem: Japanese scientists have already developed a commercially available blue carnation through the use of genetic technology. This development happened in a country where the opposition to GMOs is serious.

Engineering plants with human genes, which has been done with hemoglobin genes, seems much less acceptable. Plants that make human blood components, like hemoglobin, could help save lives, but this remains controversial. Still, the prospects offered by a new technology seem infinite, but not all prospects necessarily materialize.

Plant biotechnology is here to stay. The applications will be debated thoroughly before they are implemented. I believe that many of the questions posed today will have evaporated in the not too distant future. I predict that the era of *designer genes* (this term has been used before; I did not invent it) is near. These genes will not be bacterial, animal, or plant. They will be created entirely in the test tube and designed according to need. As we have seen, the DNA base sequence of a gene determines entirely the amino acid sequence of the protein for which it codes. Progress is being made in trying to define the biological properties of a protein based on its amino acid sequence alone. Once this can be done routinely, we can imagine that a protein with desirable properties could be computer-generated and its gene synthesized in the lab. By the time this will be feasible, the synthetic gene coding for the computer-generated protein will be introduced into plants at a predetermined location in the DNA, using enhanced gene transfer and targeting techniques. The engineered plant will then be transgenic for a synthetic gene, not one that has been isolated from a fungus, a jellyfish, or most controversially, a human being.

There is an alternative. Genetic engineering of food plants could be completely or temporarily abandoned and replaced by *pharming*, the process by which plants are engineered for the production of pharmaceuticals and vaccines. We already have seen that insulin and human growth hormone are produced by genetically engineered bacteria. Why not extend this concept to plants? In a similar argument, it is possible that one day, polluting petrochemical factories will be replaced by fields of engineered plants that synthesize biodegradable plastics.

At this point, it is impossible to predict the future of plant biotechnology because of the multitude of nonscientific issues that surrounds it. Nevertheless, I firmly believe that many applications of plant genetic engineering will be with us in the near decades.

Humans seem to have evolved in such a way that we cannot help but develop technologies. Practically all technologies have a dark side. For example, the invention of the wheel greatly facilitated transportation of people and goods. It also quickly led to the invention of the war chariot and, later on, the tank. The discovery of nuclear energy made possible the development of nuclear weapons but also made possible—for better or for worse—the establishment of nuclear power stations that provide us with inexpensive electricity. Antibiotics were truly miracle cures for several decades, until their overuse enriched pathogenic bacterial populations for resistant strains. And, as we know, computers can be used for nefarious purposes. Would the world have been better off *without* the wheel, nuclear energy, antibiotics, and computers? It is up to us to make sure that given time, a solid public understanding, and critique of the new genetic technology, we will put to use biotechnology that will benefit all of humankind.[1]

Appendix 1: Plasmids

The material presented in the appendices is intended for readers who want additional enrichment information on the topics discussed in the main text. They include scientific details on genetic engineering and its regulatory rules and more information on the green revolution that took place in the 1960s, as well as its origin and consequences.

In addition to their chromosomal DNA, many bacterial species also harbor smaller pieces of circular DNA called plasmids. Plasmid DNA does not exist in animals and plants. Bacterial plasmids occupy a large range of sizes, from very small (about 2,000 base pairs) to very large (about 400,000 base pairs). For reference, bacterial chromosomes contain from 1.5 million to 4.5 million base pairs. Most of the time, the genes carried by plasmids are not necessary for the survival of the bacteria containing them. Examples of genes present in plasmids include antibiotic resistance, detoxification of heavy metal (for instance, mercury, cadmium) salts, degradation of chlorinated organic chemicals (such as some herbicides and compounds used in dry cleaning), and toxin genes in some pathogenic bacteria.

Two types of plasmids are agronomically important: pTi found in *Agrobacterium* and the *nif* (for *ni*trogen *fi*xation) plasmid present in a class of soil bacteria called *Rhizobium*. Interestingly, *Rhizobium* species live in symbiosis with roots of leguminous plants such as peas and beans. There, the bacteria direct some root cells to form nodules in which they proliferate and start fixing atmospheric nitrogen. This nitrogen gas is eventually converted into ammonia and used by the plants to synthesize amino acids. In that sense, *Rhizobium* acts as a sort of natural fertilizer for legumes. Genetic engineers are interested in integrating the *nif* genes into the genomes of crop plants, in order to make them use gaseous nitrogen directly and reduce their dependence on chemical fertilizers. This has not yet been achieved.

In nature many types of plasmids can be exchanged among all sorts of bacterial species. Thus, bacteria initially devoid of plasmid DNA can acquire it simply by sharing a natural niche with other bacteria that do harbor plasmids. This natural transfer of plasmid DNA can have very negative consequences for humans. For example, the existence of pathogenic strains harboring antibiotic-resistance genes in hospitals is often due to plasmid transfer from innocuous bacteria to pathogenic ones. This transfer can have very undesirable consequences for hospitalized patients. On the positive side, plasmids are excellent vehicles for the cloning of DNA (see Chapter 3).

Plasmid DNA can be exchanged between bacterial species via two different mechanisms: transformation and conjugation. When cells die, they release their constituents, including plasmids, into the outside world. Then, given proper conditions, this plasmid DNA can be picked up by other bacterial cells that happen to be in the neighborhood. Once incorporated, the plasmid can start replicating in the new host and becomes established there. This phenomenon is called transformation. Uptake of plasmid DNA requires that the host bacterial cell be permeable to DNA, a rare occurrence in most natural settings. Thus, transformation is not the most common mechanism for plasmid acquisition. However, as we have seen, transformation is widely used in the laboratory because environmental conditions rendering bacterial cells permeable to DNA can be imposed there.

In nature, but often also in the laboratory, plasmids are much more likely to be transferred by the process of conjugation, a process akin to a form of primitive bacterial sexual exchange. In this case, the cell containing the plasmid builds a tube that allows

it to connect to a plasmidless partner. Through a series of complicated mechanisms, the plasmid can travel through that tube and
penetrate the new host. Since this effect is accompanied by replication of the plasmid, the donor cell does not lose it. Therefore,
at the end of the conjugation process, both partners now possess
a copy of the plasmid. The Ti plasmid from *Agrobacterium* is
transferred between bacterial cells in such a way in nature and in
the laboratory.

It is now known that *Agrobacterium* T-DNA transfer between
the bacterial cell and a plant cell proceeds in a very similar fashion. However, while the whole Ti plasmid is transferred in the
conjugation between two *Agrobacterium* cells, only its T-DNA
portion (about 10 percent of the whole plasmid) is normally
transferred when conjugation occurs between an *Agrobacterium*
cell and a plant cell.

Appendix 2: A Note on GM Animals and Potential Applications of Biotechnology to Humans

Even though this book is about plants, some comments on genetically engineered animals are warranted. Most humans consume animal products as well as plants. Further, the new genetic technology developed for plants may or may not one day be routinely applied to humans. While the genetic manipulation of plants raises all sorts of questions, genetically modifying animals—and especially humans—adds an entirely new ethical dimension to the problem of genetic modifications.

To date, transgenic mice, cows, pigs, sheep, goats, chickens, and Rhesus monkeys have been produced. In most cases, these were "proof-of-principle" experiments aimed at determining the conditions under which these engineered animals could be created. In other words, the transgenes used were easily detectable marker genes, not genes destined to enhance the quality of meat products or milk, for example.

Three techniques are used to generate transgenic animals. The first one consists of injecting a transgene by means of a very fine syringe into a fertilized egg. The transgene is usually cloned in a bacterial plasmid or in DNA isolated from a disabled virus. The injected transgene ends up integrated in chromosomal DNA, much as in the case of plants. This egg is then implanted into the uterus of a surrogate mother where it develops. Microinjection of

DNA into eggs, even though it was used first to produce transgenic animals, is tricky, and the success rate is rather low.

The second technique consists of introducing a transgene, usually by electroporation, into a special class of cells, called *embryonic
stem cells*. As their name indicates, these cells are isolated from
embryos. Further, they can be cultivated in the laboratory while
keeping their remarkable property of being *pluripotent*. This trait
means that these cells are capable of differentiating into other
types of cells. Given proper conditions, embryonic stem cells can
be transformed into brain, germ, gut, skin, and other cells. Once
embryonic stem cells have incorporated a transgene, they are
injected into a multicellular embryo called a *blastocyst*. Blastocysts
are much more resilient than eggs and tolerate injections much
better. Once inside the blastocyst, the engineered embryonic stem
cells become an integral part of the developing blastocyst and lead
to the formation of a transgenic embryo. The latter is then introduced into a surrogate mother, as detailed above.

Finally, a third technique uses an unfertilized egg from which
all chromosomes have been removed. This is done by sucking out
the nucleus of the egg with a fine syringe. At this stage, the egg is
said to be enucleated (without a nucleus). The enucleated egg is
then fused, via an electric shock, with a diploid cell isolated from
the body of another animal. These donor cells are cultivated in the
lab and are made transgenic with a gene of interest prior to fusion
with the enucleated egg. The egg, which now contains the chromosomes of the cultivated transgenic cell, is also implanted into a
surrogate mother. The famous Scottish transgenic sheep, Dolly,
was produced by fusing an enucleated egg with a transgenic udder
cell from another female. This udder cell had been made transgenic in the lab by introducing a foreign gene into udder cells cultivated in a flask. Therefore, Dolly has two mothers (the donor of
the enucleated egg and that of the udder cell) but no father!

Successful pregnancies with transgenic embryos are rare, and
most fetuses abort for reasons that are not clear. What is more,

many transgenic offspring suffer from developmental problems, such as large size, kidney dysfunction, and lameness, depending on the species considered. Mice, sheep, and goats are more resilient than cows and pigs, for example. The reasons for this disparity are not at all clear but must be addressed before animal genetic engineering becomes a reality for practical purposes. These practical purposes are at least twofold; one is the establishment of mouse models for human diseases by using the technique of gene knock-out, and the other is to turn certain animals into bioreactors, that is, living creatures producing pharmaceuticals important to humans. For example, pigs have been engineered to produce human hemoglobin, whereas sheep and goats have been engineered to produce human blood-clotting factors that could one day be used to treat hemophilia.

At this time, however, no pharmaceuticals (or enhanced meat and milk products) of transgenic animal origin are on the market. Incidentally, milk containing bovine growth hormone (practically all milk found in supermarkets) *does not* originate from GM cows. The practice of injecting growth hormone into dairy cows to enhance milk production is far from new; what is rather new is the fact that this growth hormone is now made in microorganisms that have been engineered with a cow gene to produce it. The hormone, made in large quantities, is then extracted from the GM microorganisms, purified, and administered to the cows. In other words, the genes of dairy cows producing so-called *BGH milk* have *not* been tampered with.

The ability (although only a modestly successful one) to modify animals genetically immediately raises the question of applying transgenic technologies to humans. This is an entirely different story. Even if a blind eye can be turned towards the genetic modification of plants and animals, the same cannot be done in the case of humans, because much more serious ethical considerations are involved. For example, what would be the implications of engineering humans for high intelligence, beauty, or athletic ability? Such extreme practices are not the only ones that can be contemplated in the case of human beings. Many people suffer from incurable genetic diseases that can potentially be corrected through what is called *gene therapy*. Here, the drug used to treat the disease is the correct gene itself, injected into the patient's body in the hope that it will function properly and provide the missing function. Thus, the gene is not injected into young human embryos, as is done in the case of animals. Rather, the gene is injected into specific organs, such as the liver or a muscle of a young or adult patient. Using this technique to treat hemophilia, the hereditary disease that prevents normal blood clotting, is now being seriously considered.

The problem with gene therapy is that, in order to function for prolonged periods of time, the correct gene must become part of the patient's own DNA; it must be integrated. This can be achieved by cloning the good gene into a piece of benign virus DNA known to become integrated within the human genome, such as adenovirus DNA. Unfortunately, a patient treated for a hereditary liver ailment with a gene cloned in adenovirus DNA died in September 1999. In all likelihood this catastrophe happened because the viral DNA spread into his entire body and triggered a severe immune reaction. This tragic incident shows that better vectors and more research are needed before gene therapy becomes routine.

As for human cloning, although it is an idea repugnant to many, it may soon become a reality. Scientists in the United States and Italy have made public their intention to clone volunteers (selected from infertile couples) by using the technique that created Dolly, the cloned sheep. Given the high frequencies of spontaneous abortion and fetus malformation observed in cloned animals, human cloning is a very risky proposition at present. It should be noted that the U.S. and other governments have forbidden human cloning for both medical and ethical reasons.

Appendix 3: Regulation of Biotechnology

As we saw, official protocols aimed at regulating gene technology were initiated in 1975. Over the years, review procedures have been refined and, currently, three federal agencies are involved in the review of GM plants. These are the Food and Drug Administration (FDA), the United States Department of Agriculture (USDA), and the Environmental Protection Agency (EPA). The FDA regulates the manufacture of food, food additives, cosmetics, and drugs. It evaluates products but not the processes used to manufacture these products. Thus, the FDA concerns itself with the safety of GM plant-derived foods without considering the fact that the sources of these foods are genetically modified organisms.

The USDA has the responsibility of protecting agriculture and forestry and assesses the risks associated with the release of GMOs into the environment. Finally, the EPA has jurisdiction over activities that can potentially harm the environment. GMOs fall into that category and are thus also regulated by the EPA. For example, it was the EPA that developed new regulations covering the release of plants engineered to resist pests. One can see that there exists a certain redundancy between the activities of the USDA and those of the EPA. This can be a good thing, but redundancy also poses problems of coordination and increases the complexity of review procedures. Moreover, it is not clear at this point in time that the three agencies are prepared to face the imminent avalanche of applications for certification of new GM products. Undoubtedly, federal authorities must continue to review the reviewers, until a general consensus is reached that satisfies not only bureaucrats and scientists but also the public at large.

The core issue that federal agencies and scientists alike must address is that of risk assessment. We have seen that plants engineered for insect resistance may or may not harm other insects that are not pests for these plants. Concerns are also expressed about the potential spread of transgenes by sexual crossing between engineered crops and their weedy relatives. Further, some people think that GM plants might one day outcompete indigenous species, thereby deeply altering natural environments. These questions are difficult to address, but they are not intractable. For example, a threat to innocuous insects can be evaluated in the laboratory under realistic conditions. Similarly, the spread of transgenes can be measured in greenhouse experiments and small-scale plots. Further, the fitness of GM plants and their ability to outcompete other plants also can be measured under controlled conditions.

Finally, GM plants are not just about risk. There are also benefits associated with them, as in the case of golden rice and the use of GM plants to detoxify pollutants, for example. Therefore, a risk analysis must also be accompanied by a study of positive factors to assess an acceptable level of risk, that is, the trade-off between risk and benefit. Laboratory experiments can provide a quantitative evaluation of risk levels but cannot address the psychological and cultural impact of the way risk levels are *perceived* by the public. Only objective, dispassionate information-sharing can alleviate or confirm public fears. This is exactly what is lacking in the field of GM plants at the present time.

The production and use of GM plants is also regulated via patent-granting. It should come as no surprise that life forms can be patented; this is what has allowed the biotech industry to thrive. This phenomenon is not that old, however. Indeed, it was only in 1970 that patenting of sexually reproducing plants (via seed formation) became possible. Thus, high-quality crop plants obtained by classical breeding can now be patented. However, it was only in 1980 that the first living organism, an oil-degrading bacterium, was patented, and this occurred with much heated debate. Traditionally, patents covered products of manufacture or new compositions of matter. A living organism, it was claimed, was a product of nature and hence not patentable. Courts thus rejected the request. The twist, however, was that this bacterium had been genetically engineered. The question then went all the way up to the Supreme Court that ruled in favor of the inventor. The rationale of the justices was that "what was made by man [sic]" (the engineered bacterium did not exist in nature) was patentable, regardless of the living status of the product. Patenting of genetically modified life forms—plants, animals, and bacteria—rapidly proliferated after this decision. It should be noted that, fortunately, human beings cannot be patented yet.

Proponents of the patenting of life forms argue that this is the only way they can protect large monetary investments in research and development. They also argue that intellectual property protection in the form of patents accelerates progress in biotechnology. Patenting something does not guarantee legal ownership of a product forever, though. Patent protection expires twenty years after the application is filed unless the patent is renewed. We saw in the main text that patent issues can be very thorny and can sometimes inhibit product development rather than accelerate it.

Appendix 4: Population Growth, Agronomical Production, and GM Plants

The world's population grew from 2.7 billion in 1950 to over 6 billion in 2001. I distinctly remember the 1950 value because that was the first time I heard the word *billion* uttered by anyone (my third grade teacher, actually) and was told that this was an enormous number. These numbers also mean that in one lifetime, the world's population has more than doubled. Can this trend continue? Obviously, it can't, because the resources of planet Earth are finite; one can only fit so many people in a given space and produce so much food to keep them alive. Thus, for human life to continue on Earth, the population must stop growing at a certain point.

Population growth has been most pronounced in the twentieth century. This surge is caused by modern medicine that allows more people to live through their reproductive years and has led to lower infant mortality. Some people have even predicted the demise of the human race due to overpopulation and have called population growth a "ticking time bomb." It now seems that this dark scenario will not come to pass. There are indeed good indications that our growth rate is slowing down and that the world's population may stabilize around 12 billion, sometime around the year 2100. It is in the Third World (Asia, South America, and Africa)—which now constitutes 77 percent of the world's population—that most of the growth will occur.

In spite of a slowing population growth rate and overall better food production capacity, famines persist in many areas of the world and especially in Africa. Famines are, of course, due to the inability of people to procure food. However, what is it that causes famines—lack of food, or poverty? In other words, is food getting scarce because we have grown so numerous, or is it that food is plentiful but unaffordable by some people? The neo-Malthusian school of thought (named after a British social scientist who lived at the end of the eighteenth century) maintains that population growth leads to environmental degradation, followed by a decline in food production, itself followed by poverty. Proponents of the new genetic technology favor a neo-Malthusian approach to the analysis of hunger in the world and therefore claim that higher food production based on the use of disease-resistant GM plants is necessary. Yet, this idea that greater food production alleviates famine has been shown to be incorrect. In fact, total food production per capita has increased during the last twenty years, and, even in Africa, food production per capita has remained constant since 1960. However, poverty has increased and famines have not disappeared.

Non-Malthusian social scientists say that poverty (and hence hunger) is caused by societal factors such as lack of jobs, education, and health care. These factors also contribute to excessive population growth. And indeed, population growth is low (even negative) in the fully industrialized countries of the First World. Similarly, fertility rates have declined in China and in some parts of India (in the state of Kerala, in particular) thanks to government policies that have led to income redistribution and better economic security. These examples show that population growth—and the necessity for much enhanced food production—can be curbed if social ills such as inadequate income can be repaired. In such a scenario, high-yielding GM plants may never be necessary.

Nevertheless, statistics show that hunger is on the increase in the Third World. The bitter irony is that many of the countries where famines occur actually *export* food to other countries. Clearly, in these cases, food availability is not the culprit, and a non-Malthusian interpretation is in order. It turns out that hunger-stricken people in food-exporting Third World nations are simply *too poor* to buy food (this problem also exists in the United States). Thus, the main causes of hunger are the unequal distribution of food and absence of equal opportunity to benefit from what is called *development*. In a nutshell, development programs entertained by industrial nations consist in providing aid to Third World (developing) countries so they can reach a state of self-sustenance. In fact, in the case of Brazil, for example, development programs have made hunger worse because the main beneficiaries were part of a very small upper class while the very large underclass remained in dire poverty. Here again, unequal distribution of resources was the culprit. All too often, government corruption plays a major role in this sad state of affairs as, for example, in Zaire (now the Democratic Republic of Congo) under the rule of President Mobutu. Thus, increased agricultural productivity is not the key to eradicating hunger; eradication of poverty must occur first.

Nonetheless, unequal as it is, development in Third World countries does take place and results in better wages for some. These people then gain better access to food. What kind of food? Most diets in the world center on complex carbohydrates from rice, wheat, manioc, yams, or taro accompanied by spices and vegetables. Meat and fish are generally at the periphery of the meal. In developed countries, and especially in the United States, meat or fish are core elements of the meal and are surrounded by token vegetables and rice or potatoes. Interestingly, it has been observed that development and higher wages result in increased consumption of meat and animal-derived products (such as milk) in Third World countries. This trend will have an impact on general food availability, because animals convert their feed very inefficiently and are thus an inefficient source of protein, as compared to plants. For example, 80 percent of the grain produced in the United States is fed to livestock. Add to that the fact that fodder crops are grown on land that could be used to grow plants for human consumption, that enormous water resources are needed to keep livestock healthy, and that animal husbandry generates pollution (manure, methane). With all this, one sees the problems associated with cultures shifting from a plant-based diet to an animal product-based one. Yet, this is what seems to be happening in developing countries.

Taking into consideration other problems, such as desertification, salinization, and general soil degradation, it may well be that GM plants will one day be necessary to satisfy our needs. Ironically, the use of GM plants may never be dictated by a population explosion, it may be dictated by changing mores and the degradation of the environment.

Proponents of GM plants in agriculture often refer to the green revolution and its benefits. They usually do so without a critical assessment and do not explain its origin and achievements. What follows is how the green revolution worked. In the early 1960s, the U.S. plant breeder Norman Borlaug (future Nobel Peace Prize winner) started an extensive classical breeding program (there were no GM plants in those days, of course) to improve yields in wheat. For this, he used semidwarf wheat plants that could more easily withstand the added weight of more numerous kernels without lodging (being beaten down flat). Yields did increase significantly, and a similar program was started a few years later with rice. This program was also a success. As a result, food production in the form of wheat and rice grain increased quite significantly. This was seen as a great achievement. What was not realized immediately was that improvements of yields were accompanied by a decrease in sustainability. Indeed, the new high-yielding varieties required more fertilizers and tight water control that small farmers could not afford. It also promoted monoculture (the extensive culture of a single crop) and thus decreased crop rotation, an essential factor in soil recovery. As a result, the green revolution provided benefits for a few better-off farmers, but increased landlessness and poverty for most. In addition, it marginalized Third World women because they traditionally lacked the capital to purchase fertilizers needed by the new plants. In the pre-green revolution times, women played a much greater role in agricultural production.

Further, the green revolution decreased genetic diversity found in crop plants, because it totally focused on the use of high-yielding varieties. As a result, gene banks containing older but potentially critically useful plant varieties had to be created. On the bright side, these breeding efforts were not under the control of profit-making companies, they were coordinated by a number of international plant breeding institutes supported by governments and institutions like the Ford and Rockefeller foundations. All these institutes still exist today, and, hopefully, they will be able to repeat the roles they have played in the past.

As is true of all global human efforts, the green revolution had positive and negative aspects. One blatant mistake of the green revolution was that it completely ignored Africa, focusing on Asia and South America. This neglect must end. Further, the green revolution used a top-down approach, without much concern for the little farmer and local needs. This practice must change also; the world needs a bottom-up strategy where breeding (and perhaps GM) programs are adapted to complex cropping systems already in place.

In conclusion, I do not believe that GM crops can bring the end of hunger in the world. Before making brash announcements about food shortage in the Third World, biotech companies should analyze the socioeconomic factors that promote poverty. In addition, these companies should avoid aggravating already trying circumstances by disrupting cropping systems that work (albeit with less than maximum efficiency) and possibly playing the game of corrupt government authorities who divert profits into their own satchels.

Notes

Chapter 1

1. The so-called Hardy-Weinberg law of population genetics, which allows one to calculate gene frequencies, suffers so many exceptions that it cannot really be called a *law*. At best, it is a simplified model.

2. The law of independent assortment describes the distribution of phenotypes in the offspring of a cross in which two or more genes located on different chromosomes determine these phenotypes.

Chapter 2

1. The book recounts how Watson, a young American postdoctoral scientist, joined forces with Crick, a somewhat older British physicist recently converted to biology. Their work took place at the University of Cambridge, where they once met with Chargaff before their model was a reality. Chargaff was thoroughly unimpressed "by dark horses trying to win the race" [the race to solve the structure of DNA], as Watson put it, and Crick's inability to remember the chemical formulas of the four bases. Watson's book has ruffled many feathers and remains a delightful read.

2. It is remarkable that Watson and Crick achieved their tour de force without conducting a single experiment at the bench. As in Mendel's case, their brains were their tools (the shop workers who manufactured parts for their DNA models used real tools, though). However, in contrast to them, Mendel must have gotten his hands dirty in the garden.

Chapter 3

1. The term *cloning* has more than one meaning. In the context of this book, I am referring to *gene cloning*, the isolation and multiplication of a single or a few genes from the totality of an organism's DNA. Another use of the word *cloning* refers to our ability to generate genetically identical organisms (identical multiplets) by transferring the nuclei (the cellular bodies containing a cell's DNA) of cells to eggs that have been surgically rid of their own nucleus. In the animal world, this feat has been achieved in a number of species such as frogs, sheep, cows, pigs, goats, and mice. There are projects aimed at cloning cats, dogs, chickens, rabbits, and nonhuman primates. In plants, genetically identical offspring are often easy to produce through the propagation of cut-

tings, tubers, and even single cells that can be coaxed to regenerate whole plants. Examples of plant clones are bananas, raspberries, and the Russett Burbank potato.

2. It may not have escaped the attention of the reader that the plasmid pSC101 is named after the initials of its creator, Stanley Cohen. This silly practice continued for many years after its inception. Anybody who generated or isolated a new plasmid was eager to give it his or her initials. I was tempted to do the same thing in 1976 after I had cloned pieces of barley DNA. To my dismay, my own initials, PL, had already been usurped by another scientist whose first and last names started with the same letters! I then decided to call my plasmids pHov, for *Hordeum vulgare*, the Latin name for barley. Fortunately, this self-centered practice largely stopped in the 1980s (but not early enough to prevent two of my then graduate students and a postdoc from calling their own plasmids pEP, pCP, and pBX. *Vanitas vanitatum*!

3. Another technique, the polymerase chain reaction (PCR), invented in 1986, greatly simplifies gene cloning. This invention earned its author, Kary Mullis, a self-confessed avid surfer, a Nobel Prize in 1993. The technique relies on the fact that hydrogen bonds, which hold the two strands of DNA together, can be broken simply by heating a solution of DNA in water close to the boiling point. The DNA thus becomes single-stranded. Short primers—short pieces of single-stranded DNA that can be synthesized in the laboratory and whose base sequences allow formation of a double-helical structure with the experimental DNA—are then added and the temperature is lowered. This allows the primers to bind to the experimental DNA by forming hydrogen bonds with it. At this point, the building blocks of DNA (the nucleotides) are added, together with the enzyme DNA polymerase, whose function it is to make DNA by linking nucleotides together. The polymerase binds to the primers and synthesizes the DNA's missing strand, using the nucleotides as substrates, until a full double-stranded structure is generated. Thus, after the first heating and cooling cycle is completed, one original double-stranded DNA molecule has been copied once and there are now two identical copies of this molecule. The cycle is repeated by heating again, cooling, and allowing another round of replication. There are now four new copies of the DNA molecule. After a few hours, there are millions of copies of the original DNA molecule because the process is exponential. Essentially, this procedure consists in replicating DNA in the test tube rather than in *E. coli*. This technique is much faster than conventional cloning and involves fewer manipulations. Its disadvantage is that the sequence of the gene to be amplified (multiplied) must be known in order to manufacture the specific primers needed for the reaction. Obviously, the sequences of all possible genes from all possible organisms are not known completely, meaning that conventional cloning in *E. coli* remains an important technique. PCR is such a simple procedure that amplifying DNA with it can be done in your kitchen (see the July 2000 issue of *Scientific American*)!

4. People often wonder why scientists have spent so much time and money figuring out the DNA sequence of a worm. There is a perfectly reasonable answer: *C. elegans* is used as a model system because it is completely transparent (making all its cells visible under the microscope) and the origin and development of its precisely 959 cells in the female and exactly 1,031 cells in the male are known. It also possesses a primitive nervous system and differentiated organs such as a mouth, an intestine, an anus, and sex organs. This organism is thus ideal for the study of simple behavior and organ differen-

tiation, from the fertilized egg to the adult stage. *C. elegans* is also transformable by injection of DNA.

5. The reader now understands that genetically engineering any organism is not just a matter of "cloning that gene and throwing it in there." Not only must the gene(s) be introduced into cells (in animals, this introduction can be done by microinjection into a fertilized egg), the gene(s) must also be manipulated beforehand to ensure proper expression in the new host.

Chapter 4

1. I was working as a new Ph.D. in the Ledoux laboratory at the time these great controversies exploded. I had been unable to duplicate my boss's results, which he blamed on my incompetence and arrogance. It is ironic that many of the experiments that disproved Ledoux's interpretation of DNA integration in plants were conducted in his own laboratory. These experiments were performed by visiting researchers—three from the United States, Yasuo Hotta, Andy Kleinhofs, and Clarence Kado; one from Canada, Ram Behki; plus myself. All guests visited this lab in rapid succession and even overlapped. They were very professional researchers who did not take lightly personal attacks and were impervious to European grandstanding. They simply wanted to get to the bottom of this controversy, and they did. It is interesting to note that, as per regulations of the Belgian Nuclear Center, all visitors, regardless of their academic standing, were considered trainees. Trainees in the lab were given a small wooden table without a drawer as a desk. This relegated them to the status of technician without a college degree. This also particularly irritated Andy Kleinhofs, who kept muttering, "*I* am training *you* guys."

2. It is sometimes very difficult and time-consuming to prove that something does not happen. Researchers reporting positive results (such as integration of bacterial DNA in seedlings or biological effect of foreign DNA) can always claim that inability to reproduce their results is due to the incompetence of others, their use of the wrong material, or a myriad of other reasons. One of the best ways to verify bold new claims is to spend a significant amount of time in the laboratory from which the claim originated. Alternatively, procuring experimental material from that lab can help, but this effort is more difficult because challenged researchers may not be willing to cooperate.

3. Readers would perhaps like to know who won the race, that is, who first obtained bona fide transgenic plants. This question is particularly hard to answer. Of the three studies published independently by the three competing groups, two came out in 1983 and one came out in 1984. However, all three groups announced their results at the same scientific meeting held in 1983. Also, the three labs used different experimental approaches, none of which is still in use today to produce transgenic plants. I prefer to see these three reports as complementary rather than competing. The transgenic plants made in these labs were of no practical value. The articles clearly demonstrated, however, that plant genetic engineering was feasible, and this advance was their great merit. Competition between the Monsanto, Schell/Van Montagu, and Chilton teams was intense and not always totally dispassionate. The Chilton team was the smallest and hence the most efficient. The Ghent and Monsanto teams were very large, especially the first one since Jef Schell for a while headed a Belgian and a German team number-

ing well over twenty people altogether. For the sake of justice and objectivity, I have cited all three articles in the bibliography. The reader will have noticed that the University of Leiden lab did not make it among the "winners." This group simply could not compete with the others and turned its attention to more peripheral, although important, questions of plant transformation. As for the personalities of some of the protagonists, Mary-Dell Chilton was much feared at scientific meetings; her extraordinary ability to detect nonsense in a speaker has destroyed many an ego. Jef Schell was much less intense, perhaps because he loved sailing. He has even been accused by some of simply being an excellent lab manager, no more. I presume deep jealousy may have motivated this comment.

4. At this point I cannot resist making a few comments about the country of Belgium, where I was born and educated. As we saw in this chapter, the first (and ill-fated) attempts at plant transgenesis were made there. Then, crown gall research, which culminated in successful plant genetic engineering, took a turn for the better in the same country with the Schell/Van Montagu contributions. This is terribly strange, but no doubt coincidental for a small country that most people would not be able to locate on a world map. This notwithstanding, Julius Caesar himself stated on the first page of his "Commentaries on the Gallic War" [*De Bello Gallico*] that "Of all the peoples of Gaul, the Belgians are the bravest . . . " [*Horum omnium fortissimi sunt Belgae . . .].* What an insight! Belgians today are a genetic hodgepodge, with three official languages (French, Dutch, and German) and uneasy relations between the three linguistic communities. They also have a complicated history. After 400 years of Roman occupation, the descendants of the Belgae were overtaken by the Franks. Other invaders came and went over the centuries, including the French, the Burgundians (vassals of the French court), the Spaniards, the Austrians, the French again, the Dutch, and twice the Germans. Some famous Belgians are: Pierre-Paul Rubens, Piet Breughel, and René Magritte (painters), Georges Simenon and Emile Verharen (writers), Jacques Brel (singer), Eddy Merckx (bicycle racer), Ilya Prigogine, Georges Lemaître, and Christian de Duve (scientists), Jean-Claude Vandamme (actor), Josquin des Prés, Eugène Ysaye, and César Franck (classical music composers), Tintin (cartoon character), Godiva (chocolate), Trappist (beer), Filet américain (raw ground beef mixed with an egg yolk and served with capers, pearl onions, sliced pickles, and french fries) and last but not least, Charlemagne (emperor).

Chapter 5

1. My former graduate student at Washington State University, Ray Sheehy, was instrumental in the research that led to the creation of FlavrSavr®. He lost his job when Calgene folded, a harsh punishment for having contributed to the production of the first transgenic commercial plant. The antisense technology that he used consisted in cloning the tomato polygalacturonase gene in the reverse (called antisense) orientation. When a gene, cloned backwards, is introduced by transformation in an organism, it inhibits the expression of the resident (sense) gene.

2. This research team was under the direction of Ingo Potrykus, mentioned in Chapter 4 for his discovery of plant genetic engineering by direct gene transfer. The Swiss portion of the team is housed in the Swiss Federal Institute of Technology, Zurich, where Einstein was a student and later a professor.

3. The reader may think that scientists interested in plant hybrid vigor and male sterility spend a lot of time in the field and have to scrape mud off their boots in order to be presentable. Quite the contrary. Bob Goldberg, the University of California, Los Angeles, researcher who collaborated on this male sterility study, was always impeccably dressed and would not have enjoyed mud on his shoes. Rumor has it that he has fashion and speech advisers to prep him before his trips to scientific conferences. Surely, colleague-envy must have been responsible for this gossip.

Chapter 6

1. In 1985 my laboratory received federal funding to explore the possibility of generating plants resistant to the herbicide 2,4-D. We cloned and sequenced five genes from a soil bacterium, *Alcaligenes eutrophus*, that can degrade 2,4-D all the way down to a useful metabolite, succinate. Much of this work was done in collaboration with the laboratory of Milton Gordon at the University of Washington. We then realized that transferring these bacterial genes to plants caused one major problem: One of the intermediates in the degradation of 2,4-D strongly bound to insoluble plant cellular structures and would probably be protected from further degradation by the foreign bacterial enzymes. It also turns out that this intermediate degradation product is a suspected human carcinogen (2,4-D itself is not). Thus, fearing that such transgenic plants could turn a low toxicity herbicide into a stable possible carcinogen, we abandoned research on this project. A few years ago, a biotech company planned to produce potato plants resistant to 2,4-D, using the same bacterial genes we used. I hope they also abandoned this project. At any rate, this experience stresses the need for stringent toxicity tests and the awareness of possible, yet unpredictable, reactions between plant components and herbicide degradation products.

Chapter 7

1. I am an academic scientist and have been guilty so far of the same silent attitude. I have occasionally spoken with corporate scientists about genetic engineering, but I decided many years ago to avoid any collusion with industry. Thus, I feel free to criticize academia and industry equally.

2. Not everybody in the British royal family agrees with the Prince of Wales. His father, Prince Philip of Edinburgh, sees plant biotechnology as the equivalent of breeding race horses, thus implicitly agreeing with the fact that plant genetic engineering is the same as old-fashioned plant breeding. Charles's sister Princess Anne sees nothing basically wrong with plant biotechnology. The queen, if she has an opinion, has not made it public.

3. I have spoken about this with Indian researchers visiting our Washington State University campus. They are very angry about these patent issues and feel defenseless against what they see as Western imperialism. . . again. This shows that plant biotechnology reaches much, much farther than just introducing foreign genes into plants. In addition to safety, ethical, and emotional aspects, plant biotechnology also raises questions of international politics. On the other hand, China already produces eight varieties of transgenic cotton, three varieties of transgenic sweet pepper, and one variety of transgenic tomato. In addition, the Chinese have released into the environment trans-

genic maize, potato, rice, poplar, and tobacco and are conducting field trials with trans- genic peanut, cassava, Chinese cabbage, and sweet potato. This information was pro- vided, with authorization from the Chinese Ministry of Agriculture, in the form of a let- ter to *Science* on November 24, 2000, by Bao-Hong Zhang, a member of the Cotton Research Institute of China. The scant information on transgenes used in China and the perception of GM foods there did not originate from Mr. Zhang. In fact, he never answered my e-mail request for details on GM food plants in China.

Epilogue

1. It is probably an ultimate and sour irony that the Taliban, the extremist Islamic movement that controlled most of Afghanistan for many years, declared its interest in GM cotton to the world in March 2001. These people have destroyed centuries-old statues of the Buddha and many other non-Islamic historic artifacts because they con- sidered them offensive to their faith. These same people denied education to women, and yet they were ready to embrace the use of GM cotton in their fields!

Glossary

Agrobacterium-mediated gene transfer   the act of using the soil bacterium *Agrobacterium tumefaciens* to introduce foreign genes into plants.

allele   alternative form of a gene. A single gene can have multiple alleles. For example, a hypothetical flowering plant could have a single gene determining color, but coming in several variants such as red, yellow, or blue.

amino acid   building block of proteins. There are twenty natural amino acids used by all living organisms to make proteins.

avirulence   absence of virulence.

bacteriophage   a bacterial virus.

base   building block of DNA and RNA. The bases in DNA are adenine, cytosine, guanine, and thymine. They are adenine, cytosine, guanine, and uracil in RNA.

biolistics   technique used to introduce DNA into plant cells via particle bombardment.

biotechnology   the science and art of genetically modifying organisms by directly altering their DNA.

chimeric   said of a gene that contains a promoter from one source (such as a plant) and a coding sequence from another (such as a bacterium).

chromosome   a stainable body visible in dividing cells and containing its DNA.

cloning	this term has several meanings that unfortunately can be easily mixed and confused. In the context of this book, cloning means *gene cloning,* that is, isolation, multiplication, and purification of single genes or small groups of genes. Cloning also refers to cell or organism cloning, such as in the case of Dolly the sheep. This type of cloning involves transfer of *all* the genes of an organism to a recipient cell. In the case of mammals, full genome transfer can be achieved by injection of a cell nucleus into a host such as a mammalian egg. This type of cloning is trivial in plants because many plant species can be regenerated from single cells, unlike animals. For example, a single carrot cell (among many other examples) can regenerate a whole fertile individual. This is not yet feasible in animal species, in spite of claims made in some popular novels. For example, you would be hard-pressed to produce your identical twin sister or brother from one of your skin cells. Plants do this easily.
coding sequence	the portion of a gene that contains a set of codons, each of them coding for a single amino acid.
codon	a set of three contiguous bases in RNA (three contiguous base pairs in DNA) determining one amino acid.
conjugation	bacterial mating.
cross	a mating in genetic terminology.
crown gall	a plant disease caused by the bacterium *Agrobacterium tumefaciens* and characterized by tumor formation.
cytoplasm	the portion of the cell that is not the nucleus (*see* nucleus).
diploid	a diploid cell contains two sets of chromosomes. One set originates from the father and the other from the mother. In eukaryotes, all cells of an organism are diploid except the germ cells, which are haploid.

electroporation technique that consists in subjecting cells to brief (in the millisecond range) electric pulses. This results in the formation of resealable pores in the cell membrane.

enzyme a biological catalyst necessary for metabolic reactions to occur at their natural rate. Most enzymes are proteins, some are RNA.

eukaryote a cell possessing a nucleus in which the DNA is concentrated.

exon part of a eukaryotic gene that contains amino acid-coding codons.

fatty acidsGM a class of lipids, fats, present in all living organisms.

GM genetically modified.

GMO genetically modified organism.

gene the unit of inheritance. Genes are made of DNA.

genome the set of all genes present in a cell or organism.

genotype the type of alleles or genes in a cell or organism.

genus (pl. genera) a class of living organisms given a Latin name, such as *Homo* (humans), *Canis* (dogs), *Petunia* (petunias), *Agrobacterium* (soil bacteria), *Columba* (pigeons), and *Mus* (mice). A genus is usually subdivided into various species.

haploid a haploid cell contains a single set of chromosomes. In eukaryotes, only gametes (germ cells) are haploid.

heterozygote an individual containing two different alleles of one or several genes.

homology the degree of identity, at the level of the DNA sequence, of two or more genes.

homozygote an individual containing two identical alleles of one or several genes.

Human Genome Project a multinational program whose goal is to sequence the totality of the human DNA, all 3 billion base pairs of it. This project will eventually allow the identification of all genes contained in human DNA.

intron the portion of a eukaryotic gene that does not contain amino acid-coding codons. Introns separate exons in typical eukaryotic genes.

in vitro in the test tube.

kingdom a main division in which natural organisms are classified. There are five kingdoms of life: Animalia (eukaryotes), Planta (eukaryotes), Fungi (eukaryotes), Bacteria (prokaryotes), and Archaea (also prokaryotes).

mutation an alteration of the DNA sequence of a gene.

nucleus region of the eukaryotic cell where the DNA is concentrated.

opine unusual amino acid produced by crown gall tumors.

overexpression the action of cloning a gene under the control of a strong promoter and having it expressed at high levels in a transgenic organism.

pathogen a microorganism or virus capable of infecting a multicellular organism and causing disease.

PCR *see* polymerase chain reaction.

phenotype the biological properties displayed by a cell or organism. Examples in humans include eye color, hair color, and so on. Many, but not all, phenotypic characteristics are under the control of the genotype.

phytohormone plant hormone necessary for the plant's growth and development.

phytoremediation the action of cleaning up a contaminated environment through the use of plants.

plasmid a piece of circular, double-helical DNA that coexists with the main chromosome in many bacterial species.

polymerase chain reaction (PCR) a technique allowing the multiplication of DNA molecules in the test tube. PCR can be used as an alternative to cloning in *E. coli*.

prokaryote a cell devoid of nucleus. Bacteria are prokaryotes.

promoter the region of a gene that precedes the coding sequence. The promoter is absolutely required for RNA polymerase binding and thus transcription.

protoplast plant cell rid of its cellulosic cell wall by enzymatic digestion.

pTi the tumor-inducing plasmid of *Agrobacterium tumefaciens*.

recombinant DNA a hybrid DNA molecule, made in vitro, containing some DNA from one organism and some DNA from another.

restriction enzyme or restriction endonuclease a bacterial enzyme able to cut a piece of double-helical DNA at a particular base sequence.

species a subdivision of a genus, also characterized by a Latin name. By definition, members of a species are able to interbreed. Examples are *Homo* (genus) *sapiens* (species), humans; *Canis* (genus) *familiaris* (species), dogs; *Petunia* (genus) *hybrida* (species), petunias; *Agrobacterium* (genus) *tumefaciens* (species), soil bacteria; *Columba* (genus) *livia* (species), pigeons; and *Mus* (genus) *musculus* (species), house mice. The genus *Homo* currently has only one species, *sapiens*, you the reader.

symbiosis the action of living together in close contact.

T-DNA the portion of pTi that is transferred to plant cells during crown gall formation.

terminator a sequence immediately following the coding sequence of a gene and determining transcription stop.

transcription synthesis of an RNA copy of a DNA gene.

transformation the act of introducing DNA into a recipient cell.

transgenesis the act of creating organisms containing foreign genes through DNA transfer by transformation.

translation synthesis of proteins using an RNA template, itself a copy of a DNA gene.

zygote a one-celled embryo produced by the union of a male and a female gamete.

References

Chapter 1

Mendel, G. 1959. [Original 1865]. Experiments in plant hybridization. In J. A. Peters, ed., *Classic Papers in Genetics*, pp. 1–20, Englewood Cliffs, NJ: Prentice-Hall.

Morgan, T. H. 1910. Sex limited inheritance in *Drosophila*. *Science* 32:120–122.

Sturtevant, A. H. 1913. The linear arrangement of six sex-linked factors in *Drosophila*, as shown by their mode of association. *Journal of Experimental Zoology* 14:43–59.

Chapter 2

Avery, O. T., C. M. MacLeod, and M. McCarty. 1944. Studies on the chemical nature of the substance inducing transformation of pneumococcal types: Induction of transformation by a deoxyribonucleic acid fraction isolated from pneumococcus type III. *Journal of Experimental Medicine* 79:137–158.

Watson, J. D. 1968. *The Double Helix*. New York: Mentor.

Watson, J. D., and F. H. C. Crick. 1953. A structure for deoxyribose nucleic acid. *Nature* 171:737–738.

Chapter 3

Alcamo, I. E. 1996. *DNA Technology: The Awesome Skill*. Dubuque, IA: Wm. C. Brown Publishers.

Brooker, R. J. 1999. *Genetics: Analysis and Principles*. Menlo Park, CA: Benjamin/Cummings.

Carlson, S. 2000. PCR at home. *Scientific American* 283:102–103.

Cohen, S. 1975. The manipulation of genes. *Scientific American* 233:24–33.

Cohen, S., A. Chang, H. Boyer, and R. Helling. 1973. Construction of biologically functional plasmids in vitro. *Proceedings of the National Academy of Sciences USA* 70:3240–3244.

Gilbert, W., and L. Villa-Komaroff. 1980. Useful proteins from recombinant bacteria. *Scientific American* 242:74–94.

Luria, S. E. 1970. The recognition of DNA in bacteria. *Scientific American* 222:88–102.

Mullis, K. B. 1990. The unusual origin of the polymerase chain reaction. *Scientific American* 262:56–65.

Sanger, F., G. M. Air, B. G. Barrell, N. L. Brown, A. R. Coulson, J. C. Fiddes, C. A. Hutchison, P. M. Slocombe, and M. Smith. 1977. Nucleotide sequence of the φX174 DNA. *Nature* 265:687–695.

Weaver, R. F., and P. W. Hedrick. 1991. *Basic Genetics*. Dubuque, IA: Wm. C. Brown Publishers.

Chapter 4

Barton, K. A., A. N. Binns, A. J. M. Matzke, and M.-D. Chilton. 1983. Regeneration of intact tobacco plants containing full length copies of genetically engineered T-DNA, and transmission to R1 progeny. *Cell* 32:1033–1043.

Caplan, A., L. Herrera-Estrella, D. Inzé, E. Van Haute, M. Van Montagu, J. Schell, and P. Zambryski. 1983. Introduction of genetic material into plant cells. *Science* 222:815–821.

Chilton, M.-D. 1983. A vector for introducing new genes into plants. *Scientific American* 248:50–59.

Chilton, M.-D., M. H. Drummond, D. J. Merlo, D. Sciaky, A. L. Montoya, M. P. Gordon, and E. W. Nester. 1977. Stable incorporation of plasmid DNA into higher plant cells: the molecular basis of crown gall tumorigenesis. *Cell* 11:263–271.

Hess, D. 1980. Investigations on the intra- and interspecific transfer of anthocyanin genes using pollen as vector. *Zeitschrift für Pflanzenphysiologie* 98:321–327.

Horsch, R. B., R. T. Fraley, S. G. Rogers, P. R. Sanders, A. Lloyd, and N. Hoffman. 1984. Inheritance of functional foreign genes in plants. *Science* 223:496–498.

Klein, T. M., E. D. Wolf, R. Wu, and J. C. Sanford. 1987. High velocity microprojectiles for delivering nucleic acids into living cells. *Nature* 327:70–73.

Ledoux, L., and R. Huart. 1969. Fate of exogenous bacterial deoxyribonucleic acids in barley seedlings. *Journal of Molecular Biology* 43:243–262.

Ledoux, L., R. Huart, and M. Jacobs. 1974. DNA-mediated genetic correction of thi-amineless *Arabidopsis thaliana*. *Nature* 249:17–21.

Lurquin, P. F. 1977. Integration versus degradation of exogenous DNA in plants: An open question. In W. E. Cohn, ed., *Progress in Nucleic Acid Research and Molecular Biology*, vol. 20, pp. 161–207. San Diego: Academic Press.

———. 1997. Gene transfer by electroporation. *Molecular Biotechnology* 7:5–35.

———. 2001. *The Green Phoenix: A History of Genetically Modified Plants*. New York: Columbia University Press.

Paszkowski, J., R. D. Shillito, M. Saul, V. Mandák, T. Hohn, B. Hohn, and I. Potrykus. 1984. Direct gene transfer to plants. *EMBO Journal* 3:2717–2722.

Zaenen, I., N. Van Larebeke, H. Teuchy, M. Van Montagu, and J. Schell. 1974. Supercoiled circular DNA in crown-gall inducing *Agrobacterium* strains. *Journal of Molecular Biology* 86:109–127.

Zambryski, P., H. Joos, C. Genetello, J. Leemans, M. Van Montagu, and J. Schell. 1983. Ti plasmid vectors for the introduction of DNA into plant cells without alteration of their normal regeneration capacity. *EMBO Journal* 2:2143–2150.

Chapter 5

De Greef, W., R. Delon, M. De Block, J. Leemans, and J. Botterman. 1989. Evaluation of herbicide resistance in transgenic crops under field conditions. *Bio/Technology* 7:61–64.

Duke, S. O. 1996. *Herbicide-resistant Crops*. Boca Raton, FL: CRC Lewis Publishers.

Gasser, C. S., and R. T. Fraley. 1992. Transgenic crops. *Scientific American* 266:62–69.

Gura, T. 1999. New genes boost rice nutrients. *Science* 285:994–995.

Hinchee, M. A. W., D. V. Connor-Ward, C. A. Newell, R. E. McDonnell, S. J. Sato, C. S. Gasser, D. A. Fischhoff, D. B. Re, R. T. Fraley, and R. B. Horsch. 1988. Production of transgenic soybean plants using *Agrobacterium*-mediated DNA transfer. *Bio/Technology* 6:915–922.

Ku, M. S. B., D. Cho, U. Ranade, T.-P. Hsu, X. Li, D.-M. Jiao, J. Ehleringer, M. Miyao, and M. Matsuoka. 2000. Photosynthetic performance of transgenic rice plants overexpressing maize C4 photosynthesis enzymes. In J. E. Sheehy, P. L. Mitchell, and B. Hardy, eds., *Redesigning Rice Photosynthesis to Increase Yield*. Makati City (Philippines) and Amsterdam (The Netherlands): Elsevier Science.

Martin, G. B., S. H. Brommonschenkel, J. Chunwongse, A. Frary, M. W. Ganal, R. Spivey, T. Wu, E. D. Earle, and S. T. Tanksley. 1993. Map-based cloning of a protein kinase gene conferring disease resistance in tomato. *Science* 262:1432–1436.

Nash, J. M. 2000. Grains of hope. *Time* 156:38–46 (July 31, 2000 issue).

Ronald, P. C. 1997. Making rice disease-resistant. *Scientific American* 277:100–105.

Vaeck, M., A. Reynaerts, H. Höfte, S. Jansens, M. De Beuckeleer, C. Dean, M. Zabeau, M. Van Montagu, and J. Leemans. 1987. Transgenic plants protected from insect attack. *Nature* 328:33–37.

Wehrmann, A., A. Van Vliet, C. Opsomer, J. Botterman, and A. Schulz. 1996. The similarities of *bar* and *pat* gene products make them equally applicable for plant engineers. *Nature Biotechnology* 14:1274–1278.

Ye, X., S. Al-Babili, A. Klöti, J. Zhang, P. Lucca, P. Beyer, and I. Potrykus. 2000. Engineering the provitamin A (β-carotene) biosynthetic pathway into (carotenoidfree) rice endosperm. *Science* 287:303–305.

Chapter 6

Apse, M. P., G. S. Aharon, W. A. Snedden, and E. Blumwald. 1999. Salt tolerance conferred by overexpression of a vacuolar Na^+/H^+ antiport in *Arabidopsis*. *Science* 285:1256–1258.

de la Fuente, J. M., V. Ramirez-Rodriguez, J. L. Cabrera-Ponce, and L. Herrera-Estrella. 1997. Aluminum tolerance in transgenic plants by alteration of citrate synthesis. *Science* 276:1566–1568.

DellaPenna, D. 1999. Nutritional genomics: Manipulating plant micronutrients to improve human health. *Science* 285:375–379.

Gerngross, T. U., and S. C. Slater. 2000. How green are green plastics? *Scientific American* 283:37–41.

Haq, T. A., H. S. Mason, J. D. Clements, and C. J. Arntzen. 1995. Oral immunization with a recombinant bacterial antigen produced in transgenic plants. *Science* 268:714–716.

Kovalchuk, O., I. Kovalchuk, V. Titov, A. Arkhipov, and B. Hohn. 1999. Radiation hazard caused by the Chernobyl accident in inhabited areas of Ukraine can be monitored by transgenic plants. *Mutation Research* 446:49–55.

Langridge, W. H. R. 2000. Edible vaccines. *Scientific American* 283:66–71.

Mariani, C., M. De Beuckeleer, J. Truettner, J. Leemans, and R. B. Goldberg. 1990. Induction of male sterility in plants by a chimaeric ribonuclease gene. *Nature* 347:737–741.

Mariani, C., V. Gossele, M. De Beuckeleer, M. De Block, R. B. Goldberg, W. De Greef, and J. Leemans. 1992. A chimaeric ribonuclease-inhibitor gene restores fertility to male sterile plants. *Nature* 357:384–387.

Mazur, B., E. Krebbers, and S. Tingey. 1999. Gene discovery and product development for grain quality traits. *Science* 285:372–375.

Moffat, A. S. 1995. Exploring transgenic plants as a new vaccine source. *Science* 268:658–660.

_____. 1995. Plants as chemical factories. *Science* 268:659.

Murata, N., O. Ishizaki-Nishizawa, S. Higashi, H. Hayashi, Y. Tasaka, and I. Nishida. 1992. Genetically engineered alteration in the chilling sensitivity of plants. *Nature* 356:710–713.

Perkins, E. J., M. P. Gordon, O. Caceres, and P. F. Lurquin. 1990. Organization and sequence analysis of the 2,4-dichlorophenol hydroxylase and dichlorocatechol oxidative operons of plasmid pJP4. *Journal of Bacteriology* 172:2351–2359.

Poirier, Y., D. E. Dennis, K. Klomparins, and C. Sommerville. 1992. Polyhydroxybutyrate, a biodegradable thermoplastic, produced in transgenic plants. *Science* 256:520–522.

Thomashow, M. F. 1999. Plant cold acclimation: Freezing tolerance genes and regulatory mechanisms. *Annual Reviews of Plant Physiology and Molecular Biology*, vol. 50, pp. 571–599. Palo Alto, CA: Annual Reviews, Inc.

Chapter 7

Barnum, S. R. 1998. *Biotechnology: An Introduction*. Belmont, CA: Wadsworth Publishing Company.

Bodley, J. H. 1996. *Cultural Anthropology: Tribes, States and the Global System*. 2nd. ed. Mountain View, CA: Mayfield.

Brasher, P. 2000. Company says altered corn is very low risk, but questions persist. *The Oregonian*, Portland, OR. October 27, p.A2.

Brown, K., K. Hopkin, and S. Nemecek. 2001. Genetically modified foods: Are they safe? *Scientific American* 284:51–65.

Chrispeels, M. J., and D. E. Sadava. 1994. *Plants, Genes, and Agriculture*. Boston, MA: Jones and Bartlett Publishers.

Daniell, H. 1999. GM crops: Public perception and scientific solution. *Trends in Plant Science* 4:467–469.

Daniell, H., R. Datta, S. Verma, S. Gray, and S. Lee. 1998. Containment of herbicide resistance through genetic engineering of the chloroplast genome. *Nature Biotechnology* 16:345–348.

Duke, S. O. 1996. *Herbicide-resistant Crops*. Boca Raton, FL: CRC Lewis Publishers.

Gaskel, G., M. W. Bauer, J. Durant, and N. C. Allum. 1999. Worlds apart? The reception of genetically modified foods in Europe and the U.S. *Science* 285:384–387.

Goldman, K. A. 2000. Bioengineered food—Safety and labeling. *Science* 290:457–459.

Horvath, H., J. Huang, O. Wong, E. Kohl, T. Okita, C. G. Kannangara, and D. von Wettstein. 2000. The production of recombinant proteins in transgenic barley grains. *Proceedings of the National Academy of Sciences USA* 97:1914–1919.

Joersbo M., J. Donaldson, K. Kreiberg, S. Guldager Petersen, J. Brunstedt, and F. T. Okkels. 1998. Analysis of mannose selection used for transformation of sugar beet. *Molecular Breeding* 4:111–117.

Lappe, F. M., J. Collins, and C. Fowler. 1979. *Food First: Beyond the Myth of Scarcity*. New York: Ballantine.

Robbins, R. H. 1999. *Global problems and the culture of capitalism*. Needham Heights, MA: Allyn and Bacon.

Serageldin, I. 1999. Biotechnology and food security in the 21st century. *Science* 285:387–389.

Tsao, E. 2000. Western Family recalls products with altered corn. *The Oregonian*, Portland, OR. October 26, p. A2.

Epilogue

Moffat, A. S. 2000. Can genetically modified crops go "greener"? *Science* 290:253–254.

Ow, D. W., K. V. Wood, M. DeLuca, J. R. De Wet, D. R. Helinski, and S. H. Howell. 1986. Transient and stable expression of the firefly luciferase gene in plant cells and transgenic plants. *Science* 234:856–859.

Web sites of interest

www.agrevo.com/biotech/QA/qa_dt.htm#Anfang This web site was created by AgrEvo, the maker of Liberty® and producer of transgenic resistant canola. It offers questions and answers regarding plant biotechnology.

www.biotech-info.net This is a very comprehensive site that objectively presents the pros and cons of plant biotechnology. It provides an extensive list of technical and nontechnical references.

www.cgiar.org/biotech/rep0100/contents.htm This web site shows the articles read at the international conference "Agricultural Biotechnology and the Poor." All articles deal with biotechnology and the Third World.

www.colostate.edu/programs/lifesciences/TransgenicCrops This Colorado State University site aims at providing balanced information on existing and future transgenic crops, including a discussion of risks.

www.fda.gov/ This is the official home page of the Food and Drug Administration. It provides extensive information on the safety of genetically engineered food plants.

www.greenpeace.org/ This is the Greenpeace International web site. Its content is firmly opposed to biotechnology and its use with crop plants. This page contains many other topics (usually in the form of exposés), mostly related to the environment.

www.monsanto.com/monsanto/default.htm This is Monsanto's web site. It also gives information regarding plant biotechnology as seen from the industry's side. Monsanto is the company that released Roundup®-resistant soybean and insect-resistant corn.

www.whybiotech.com/ This is a web site generated by the Council for Biotechnology Education, an organization set up by a consortium of major biotech companies: Aventis CropScience, BASF, Dow Chemical, DuPont, Monsanto, Novartis, and Zeneca Ag Products, Inc. This home page obviously presents the industry's viewpoint.

Index

A. *tumefaciens*, 77–78, 80, 96
Acid rain, 122
Afghanistan, 192(n1)
AgrEvo, 99, 103
Agriculture, 2
Agrobacterium, 142, 143, 169, 171
Agrobacterium-mediated gene
    transfer, 77–94, 85, 100,
    102–103, 104, 108, 114, 127
Agrobacterium radiobacter, 78
Agrobacterium T-DNA, 106
Agrobacterium tumefaciens,
    73–77, 81
Agrochemicals, 160
Agronomical production, 182–186
Alar®, 98
Albino alelle, 34
Albino gene, 17
Albino mutant, 8
Alcaligenes eutrophus, 191(n1)
Alleles, 13–14, 34
Allergenicity, 148–149
Aluminum detoxification, 134
Amino acids, 29
Ammonia, 100, 102–103
Animal cloning, 187–188(n1). *See
    also* Cloning

Animal husbandry, 2
Animals, transgenic, 60–61,
    173–176
Animal viruses, 128
Anne, Princess, 191(n2)
Antibiotic resistance, 82
Antibiotic-resistance genes,
    142–144
Antibodies, 128
Antigens, 128
Antisense technology, 113,
    190(n1)
Apples, 98
Applied science, 95
Aquatic plants, 120
Arabidopsis, 68, 70, 120, 121–122,
    126, 130–131, 133, 165
Arabidopsis thaliana, 51, 67
Arntzen, Charles, 127–129
Asilomar Conference, 59
AtNHX1, 121
Australia, 160
Australopithecus, 2
Aventis CropScience, 151
Aventis Pharmaceuticals, 99
Avery, Oswald, 27–30, 38, 42–43,
    47, 65

Gordon, Milton, 78, 191(n1)
Grants, 145, 146
Grapevines, 106–107
Greenpeace, 139, 162–163
Green plastics, 133
Green revolution, 156–157,
    159–160, 184–185
Griffith, Frederick, 28

Hardy-Weinberg law of popula-
    tion genetics, 187(n1)
Harvard University, 51
Health issues. *See* Human health;
    Public health concerns
Heavy metal pollutants, 120,
    129–130, 135
Hemophilia, 176
    and transgenic animals, 175
    See also Diseases
Herbicide-resistance genes,
    146–147
Herbicide-resistant crops, 57,
    113, 137, 163, 164
Herbicide-resistant Crops
    (Duke), 163
Herbicide-resistant plants, 156
Herbicides, 96–97, 97–104, 134,
    191(n1)
Hereditary diseases, 176. *See also*
    Diseases
Heredity, 8–9, 17, 26, 28
Heredity, laws of, 4–5, 13, 17–18
Hershey, Alfred, 29
Hess, Dieter, 65–66, 69–70, 71,
    72
Hohenheim University, 65
Homo sapiens, 2

Horticulture, 165–166
Hotta, Yasuo, 189(n1)
Human cancer, 73
Human cloning, 176. *See also*
    Cloning
Human error, 150
Human genes, 166
Human genome, 34–35, 51
Human Genome Project, 51
Human growth hormone, 60
Human health, 125–129,
    148–150. *See also* Public
    health concerns
Human insulin, 60
Human medicine, 164
Human nutrition. *See* Nutrition
Humans, and heredity, 8
Humans, transgenic, 175–176
Human viruses, 128
Hungarian Academy of Sciences,
    119
Hunger, 159–160, 183. *See also*
    Famines
Hybrid seeds, 114–116
Hydrocarbon-degrading ability, of
    bacteria, 129

Ice nucleation gene, 125
Identical multiplets, 187–188(n1)
Independent assortment, law of,
    16, 187(n2)
India, 160
Indica rice, 151. *See also* Golden
    rice; Rice
Insect-resistant crop plants,
    96–97, 97, 113, 137, 146,
    147–148, 164